FACTBOOK
—OF—
SCIENCE

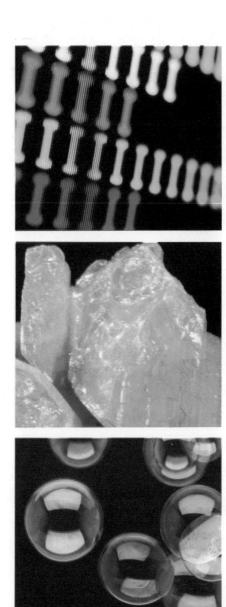

FACTBOOK
—OF—
SCIENCE

MIMOSA
·BOOKS·

This edition published in 1993 by
Mimosa Books, distributed by Outlet Book
Company, Inc., a Random House
Company, 40 Engelhard Avenue, Avenel,
New Jersey 07001.

10 9 8 7 6 5 4 3 2 1

First published in 1977 by Grisewood &
Dempsey Ltd.
Copyright © Grisewood & Dempsey Ltd.
1977, 1982, 1993

All rights reserved. No part of this
publication may be reproduced, stored in
a retrieval system or transmitted by any
means, electronic, mechanical,
photocopying or otherwise, without the
prior permission of the publisher.

ISBN 1 85698 521 0

Printed and bound in Italy

CONTENTS

Atoms and Elements	11
Universal Forces	24
Energy	26
Force and Motion	30
Relativity	40
The Electromagnetic Spectrum	43
X-rays	49
Lasers	50
The Spread of Colour	54
Radio Waves	57
Electricity and Magnetism	64
d.c. Circuits	68
Electromagnetism	71
Alternating Current	83
Radio and Television	84
Computers	96
Sound	102
Heat and Temperature	110
Putting Heat to Work	121

Colder than Cold	128
Chemicals	130
Chemical Elements	142
Iron and Steel	154
Copper	164
Nickel	167
Precious Metals	169
Tin, Lead and Zinc	174
Refractory Metals	176
Alkali Metals	178
Alkaline Earth Metals	180
The Atmosphere	182
Oxygen	186
The Rare Gases	190
Nitrogen	192
Water	194
The Periodic Table	198
A-Z of Science	200
Branches of Science	228
SI Units	229
Index	230

Atoms & Elements

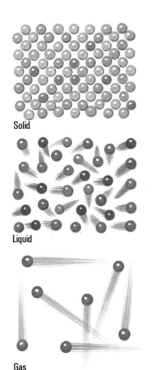

Solid

Liquid

Gas

Everything is made up entirely of atoms. Whether something is solid, liquid or gas depends merely on how closely the atoms are packed.

Left: Atoms are so small that if an atom were the size of your little finger nail, then your hand would be large enough to grasp the Earth!

Lord Rutherford put forth the modern theory of the atom in 1911. The atom is not a solid, indivisible piece of matter, but consists mainly of space. It is rather like a miniature solar system. At its centre, or nucleus, is concentrated the main mass of the atom. The nucleus consists mainly of two kinds of 'fundamental' particles – protons, which have a positive electric charge, and neutrons, which have a similar mass to the protons, but no electric charge. Circulating around the nucleus is a number of even smaller particles called electrons. Electrons have a negative electric charge. There are equal numbers of protons and electrons so that the atom as a whole is electrically neutral.

An atom is only a few hundred millionths of a centimetre across. A proton is so light that it would take 160 million million million million of them to weigh one gramme. An electron is 1800 times lighter than this.

Every element has a different number of protons in its nucleus. The number of protons is called the atomic number. Hydrogen has the simplest of all atoms, with only a single proton in its nucleus, and has the atomic number 1. The most complex element found in nature – uranium – has 92 protons in its nucleus, and thus has the atomic number 92.

All the elements except hydrogen always have neutrons in their nucleus. Often the same element has atoms with different numbers of neutrons in their nucleus. Iron, for example, consists of a mixture of different atoms with 28, 30, 31, and 32 neutrons in their nuclei. Hydrogen atoms can have none, one, or two neutrons. Such atoms are called isotopes. Among the 90 chemical elements found in nature there are more than 320 isotopes.

Radioactivity

Most of the isotopes found in nature are stable – that is, they do not change. But about 50 of them are unstable. They break down spontaneously, giving off streams of particles, or radiation, or both. Such isotopes are called radioactive. The best known radioactive elements are uranium and radium.

Three basic kinds of particles or radiation are emitted by radioactive isotopes – alpha-particles, beta-rays, and gamma-rays. Alpha-particles consist of two protons and two neutrons, and thus have a positive charge. They are identical to the nuclei of helium atoms. Beta-rays are streams of electrons.

Above: An alpha particle is emitted from the nucleus of an atom. It consists of two protons and two neutrons, and is in fact the nucleus of a helium atom.

An atom of phosphorus. The nucleus contains fifteen positively charged protons and an equal number of neutrons, which have no charge. Round this nucleus circle fifteen electrons. Electrons have a negative charge and since there is an equal number of protons and electrons the atom as a whole is electrically neutral. The electrons orbit the nucleus of the phosphorus atom in three shells – two in the inner shell, eight in the middle shell, and five in the outer shell. An electron is 1800 times as light as a proton.

Atoms and Elements

Above: A beta particle, which is an electron, is emitted when a neutron in the nucleus changes into a proton and an electron.

Gamma-rays consist not of charged particles, but of electromagnetic waves like radio waves and X-rays. They have a very short wavelength, high energy, and are extremely penetrating. They are the most dangerous of the radiations given off by radioactive elements.

When a radioactive element gives off alpha-particles, each of its atoms loses two protons from its nucleus. But it is the number of protons in the nucleus of an atom that determines what element it is. So when an atom of one element gives off alpha-particles, it changes into the atom of another element. The number of protons in the nucleus of an atom also

Henri Becquerel (1852-1908) was a French physicist who discovered radioactivity in 1896. He found that uranium compounds would fog a photographic plate, even though it had not been exposed to light. Becquerel believed that the uranium compound produced rays that could penetrate the wrapping of the plate. These rays were in fact beta rays. Becquerel showed that they were streams of electrons.

Marie Curie

Pierre Curie

Henri Becquerel

Marie Curie (1867-1934) and **Pierre Curie** (1859-1906) were French chemists. They looked for an unknown source of radio-activity in uranium minerals. After many years of hard work, they obtained a small amount of radium from tons of uranium ore. It was extremely radioactive, and produced intense alpha rays. The rays were streams of helium nuclei.

Atoms and Elements

changes when it emits beta-rays. So again that atom changes into the atom of another element. Such a change is called a transmutation.

Further transmutations will occur if the new atoms formed are again radioactive and unstable. They will continue until an atom is formed that is stable. Atoms of uranium, for example, first change into atoms of thorium. But these atoms are also radioactive, and they change into atoms of protactinium. These are also radioactive and change into atoms of radium, which are radioactive. They change eventually into atoms of lead, which is stable. So there the radioactive series ends.

Making new elements

The way radioactive elements change into other elements prompted scientists to try changing elements themselves. Lord Rutherford was the first to attempt this; in 1919 he succeeded in changing nitrogen into oxygen by bombarding nitrogen gas with a stream of alpha-particles from uranium. This is what happened on the atomic scale: an atom of nitrogen (7 protons) captured an alpha-particle (2 protons) and at the same time gave off a proton, leaving an extra proton in the nucleus. But an atom with 8 protons in the nucleus is an atom of oxygen. So nitrogen had been changed into oxygen.

Bombardment by beta-rays and neutrons will also change one element into another. A whole new series of chemical elements, not found in nature – plutonium, neptunium, americium, and so on – has been made by bombarding uranium and other elements. So far 16 artificial elements have been produced by nuclear bombardment – all of them radioactive and short-lived.

Bombardment of an element will often produce isotopes of that element or other elements

An atom's ability to enter into chemical combination with another atom is termed its valency. The valency of an element is the number of electrons it needs to gain or lose to make it stable (a total of eight electrons in its outermost shell).

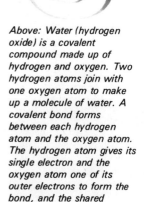

Above: Water (hydrogen oxide) is a covalent compound made up of hydrogen and oxygen. Two hydrogen atoms join with one oxygen atom to make up a molecule of water. A covalent bond forms between each hydrogen atom and the oxygen atom. The hydrogen atom gives its single electron and the oxygen atom one of its outer electrons to form the bond, and the shared electrons move around both atoms.

Atoms and Elements

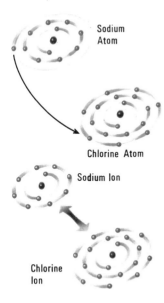

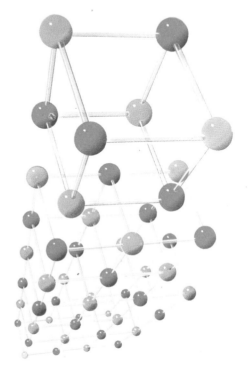

Below: Salt (sodium chloride) is made up of sodium ions (green) and chlorine ions (yellow) all joined in a regular network.

Salt (sodium chloride) is an ionic compound made up of sodium and chlorine ions. When it forms, each sodium atom loses one of its outer electrons, which goes to a chlorine atom. An ionic bond forms as the negative chlorine ion is attracted to the positive sodium ion.

that are radioactive. Such radioactive isotopes, or radio-isotopes, can be of enormous value in science, engineering, and medicine.

Scientists usually produce radio-isotopes by exposing materials to the intense radiation in a nuclear reactor. In their study of atoms they often carry out nuclear bombardment with accelerated particles. They use machines called particle accelerators – 'atom-smashers' – to do this. The general principle behind these machines is that charged particles are accelerated by applying a powerful electric charge to them so that they are strongly repelled.

Atoms and Elements

Uranium occurs in nature as a mixture of two main isotopes, which we term U-238 and U-235, the number being the number of particles in the nucleus. Although U-235 occurs only in minute amounts it is very important. When U-235 is bombarded with neutrons under suitable conditions, the atom absorbs a neutron and becomes completely unstable. It splits into two main fragments, giving off energy and, what is particularly important, two or more neutrons. We call this splitting process *fission*.

The two or more neutrons emerging from the split atom can split other uranium atoms. In turn these other fissions will yield other neutrons, which will be able to split still more atoms. A *chain reaction* occurs, with more and more neutrons being produced and more and

Enrico Fermi, the Italian scientist who was the first to produce nuclear fission and to achieve a chain reaction.

Sir Joseph John Thomson (1856–1940), a British physicist, showed in 1897 that cathode rays consist of beams of minute particles bearing negative electric charges. The particles were called electrons, a name that had already been put forward, but Thomson is considered to have discovered the electron.

Ernest Rutherford (1871–1937) was born in New Zealand but worked in Britain and Canada. He studied under J. J. Thomson and became the first man to explain the structure of the atom in 1911. Earlier experiments by Rutherford had showed unexpectedly that alpha particles (positive particles) are scattered when fired through gold foil. Rutherford reasoned that a positive nucleus within the gold atoms deflects the particles. He concluded that the atom has a positive nucleus surrounded by negative electrons. Rutherford was later knighted.

The two men went on to make other important discoveries. In the same year as Rutherford announced his discovery of the nucleus, Thomson obtained the first indications that isotopes exist. Rutherford went on to split the atom.

Atoms and Elements

> **E=mc²**
> When the total mass of the products of nuclear fission is calculated, it is found to be less than the original mass of the atom before fission. Albert Einstein's Theory of Relativity provides an explanation for this 'lost' mass. Einstein demonstrated that mass and energy are equivalent things. So the mass lost during fission reappears as energy. Einstein summed up the equivalence in the famous equation $E=mc^2$ where E = energy, m = mass, c = the speed of light. Since c is very large (186,000 miles, 300,000 km a second) E will be very large indeed, when even a little mass is 'lost'.
>
> Einstein was born in 1874 at Ulm, Germany, of German-Jewish parents. He studied mathematics and physics at Zurich and then worked in the Swiss Patent Office at Bern. In 1905 he published four papers, each of which formed the foundation of a new branch of physics. In these papers he explained the phenomenon of photo-electricity, the equivalence of mass and energy, Brownian movement, and his special theory of relativity encompassing revolutionary ideas on mass, space, gravitation, time, and motion.

more atoms being split. And a tremendous amount of energy will be released. For a chain reaction to occur, the uranium must be in a sufficiently large lump; otherwise too many neutrons escape. The smallest amount of uranium necessary is called the *critical mass*.

Nuclear reactors

An uncontrolled chain reaction gives rise to a devastating explosion, as in the atomic bomb. In a nuclear reactor a controlled chain reaction takes place. Nuclear reactors are the heart of atomic power stations. They produce heat by splitting atoms, unlike ordinary power stations which produces heat by burning fuel. This heat is then made to boil water, to drive steam turbines for generating electricity. Nuclear

Atoms and Elements

reactors are used in a similar way in nuclear submarines for raising steam to drive the turbines that turn the propellers.

The fuel used in most reactors is natural uranium – a mixture of U-238 and U-235 – sometimes enriched with extra U-235. The U-238 tends to absorb the high-speed neutrons given off by splitting U-235 atoms, but it does not absorb slow neutrons quite so readily. So in a reactor a substance called a moderator is included with the uranium, to slow the neutrons down. The U-238 will then not absorb them so easily and fission will continue.

To make sure that the rate of fission does not get out of hand, control rods are inserted in the reactor core. These rods are made out of a material such as boron and cadmium that readily absorbs neutrons. They are pushed in or out of the core as necessary. To extract the heat developed in the reactor, a cooling fluid, or coolant, is circulated through the core.

A reactor produces great amounts of heat and intense streams of neutron and gamma radiation. Neutron and gamma radiation are deadly to all forms of life in even small amounts, causing sickness, leukaemia, and ultimately death. The reactor must be surrounded with thick biological shielding of concrete and steel to prevent harmful radiation escaping. Radioactive materials are handled by remote control and stored in lead containers – lead makes an excellent shield against radiation.

Breeding Fuel

Another kind of nuclear reactor can actually produce more fuel than it consumes. This is the *fast reactor* or *breeder reactor*, which has just begun to produce useful power. It has no moderator and its fuel is highly enriched uranium or plutonium. The core is small and

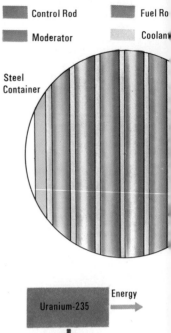

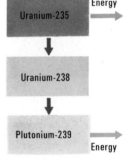

Above: In a fast reactor, fuel highly enriched in uranium-235 produces energy. Large amounts of neutrons are also given out, but a blanket of uranium-238 placed around the reactor core absorbs the neutrons, which convert it into plutonium-239. This plutonium isotope is then separated and later used as fuel in other nuclear reactors.

Atoms and Elements

The Charging Floor lies above the core of the reactor. Here, fuel rods are inserted into the core and removed after use. The control rods are operated from outside the core.

THE NUCLEAR REACTOR

The Heat Exchanger takes heat from the core. Hot coolant from the core flows around the central tube of the heat exchanger, boiling the water inside. This lowers the temperature of the coolant, which is then pumped back to the core. In some reactors, water is used as coolant and it boils in the core.

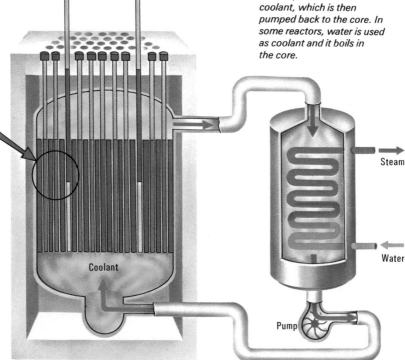

The Core of the reactor is built inside a strong steel container. It contains fuel rods made of fissile materials housed within tubes. The fuel rods produce heat as the fuel undergoes fission. The control rods are moved in and out of the core to vary the flow of neutrons in the core, controlling the rate of fission and therefore the heat produced. The rods are surrounded by a moderator, which slows the neutrons produced by the fuel so that fission proceeds steadily. Through the core flows a coolant, a liquid or a gas, which removes heat from the core.

the chain reaction proceeds rapidly, producing greater amounts of heat than with the other 'thermal' reactors. Large amounts of neutrons are produced but absorbed in a blanket of uranium-238 placed around the core. This does not cause fission in the uranium, but converts it into plutonium-239. This plutonium can later be separated and then used as a fuel for the fast reactor. In this way, the fast

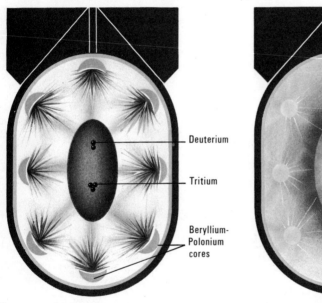

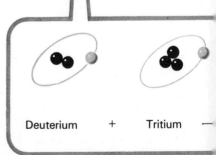

Above: The principle of the hydrogen bomb. Beryllium-polonium cores trigger plutonium explosions which produce intense heat and pressure. Under these conditions, tritium and deuterium fuse to form helium. In practice, a compound of lithium and deuterium is used instead of deuterium and tritium, and neutrons from the triggering explosions create tritium from the lithium. The fusion reaction then follows. In some bombs, neutrons from this fusion explosion then set off a third stage—the fission of uranium in the casing of the bomb.

Atoms and Elements

reactor produces fuel as it consumes it. By converting non-fissile uranium-238 into a useful fuel, the fast reactor could extend the world's reserves of nuclear fuel by sixty times!

The Hydrogen Bomb

The first thermonuclear weapon, or hydrogen bomb as it is often called, was exploded in a test by the United States in 1952. It detonated with a force of 10 megatons—equal to exploding 10 million tons of TNT or conventional high explosive. The power of this terrible weapon was 750 times greater than that of the first atomic bombs, and sufficient to wipe out any capital city. In the hydrogen bomb, an atomic-bomb trigger sets off a fusion reaction in a compound of deuterium and lithium, the lithium instantly producing tritium that reacts with the deuterium. But military scientists also thought of a way of increasing the power of the bomb by encasing it in natural uranium. The powerful neutrons produced by fusion then cause an even greater fission explosion in the uranium casing. In 1961, Russia tested the most powerful bomb ever made. It exploded with a force of 57 megatons.

A fission bomb produces a deadly cloud of radioactive fission products. Some of these products rain down on the area around the explosion, causing radiation sickness in anyone caught there—as at Hiroshima and Nagasaki. The remainder of the dust rises into the atmosphere and spreads out before returning to the ground, often in rain. This radioactivity that descends invisibly from the skies is called *fall-out* and it may enter the body. Fusion does not produce radioactive products and so a fusion bomb is 'clean' and devoid of fall-out—apart from radioisotopes produced as the casing vaporizes. But such a

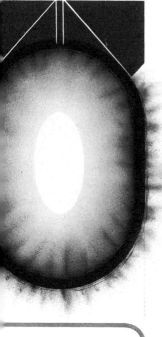

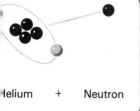

Helium + Neutron

Atoms and Elements

weapon is not really clean; it kills by the deadly radiation that it produces and death, as we have seen in Japan, may take a very long time. The uranium jacket around the hydrogen bomb produces immense amounts of fall-out and such bombs have become known as 'dirty' bombs.

Controlling Fusion

As hydrogen bomb tests began to poison the atmosphere, many scientists began to work to find ways of controlling fusion to produce useful power. The problems were immense, far greater than those in making the first fission reactor. The main one was how to heat the hydrogen isotopes to the great temperatures needed to start fusion. How could such a temperature be achieved in a machine without it vaporizing?

The main line of research has been to confine the isotopes in a magnetic field within a container. A high electric current is passed through the hydrogen to heat it up and form a plasma—a gas composed of charged atoms—while a magnetic field compresses the plasma and keeps it from touching the walls. Even if it does touch the container, vaporization does not occur because very small amounts of hydrogen are heated; the walls simply cool the plasma more than the plasma heats the walls.

To achieve fusion, more than high temperature is needed. There must be sufficient plasma present so that the nuclei will meet and fuse, and the high temperature must be produced for long enough for this to happen. The right combination of all these factors has so far proved impossible to achieve. The plasma does not remain stable and twists out of the magnetic field before sufficiently high temperatures or densities can be attained.

Atoms and Elements

Research is going on in many countries to improve the containment of plasma, but success still appears to be distant.

Fusion scientists are therefore looking at other ways of obtaining controlled thermonuclear power, and a new method looks promising. By firing laser beams or electron beams at pellets of solid deuterium and possibly tritium, fusion might occur in the pellets. The beams would heat the pellets to very high temperatures, but the heating would be so rapid that the interior of the pellets would be placed under great pressure before they flew apart. In these conditions, fusion might occur.

ATOMS IN SPACE

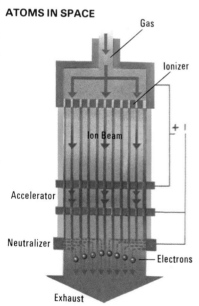

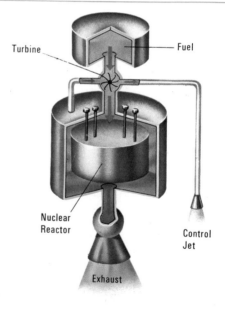

An Ion Engine may be used for space travel in the future. It is powered by electricity, obtained either by solar cells or a nuclear generator, and uses a gas as a propellant. The gas atoms pass through a charged metal screen to ionize them, and are then accelerated in an electric field. As the ion beam leaves the exhaust, its electrical charge is neutralized by adding electrons.

A Nuclear Rocket Engine would simply heat a propellant gas to make it expand and rush from the exhaust. Some of the gas would be diverted to drive a turbine to pump the fuel from its tank and to power control jets.

Universal Forces

Protons and neutrons are held together to make up the nucleus, the nucleus attracts electrons to form an atom, atoms group themselves into molecules, assemblies of molecules make up whole worlds, worlds circle in packs around stars, stars form together in millions to produce galaxies, and the galaxies in turn make up the final supergroup—the Universe. What makes these particles or bodies gather together to form a succession of groups? What holds the Universe together?

The answer is force. Four kinds of forces are

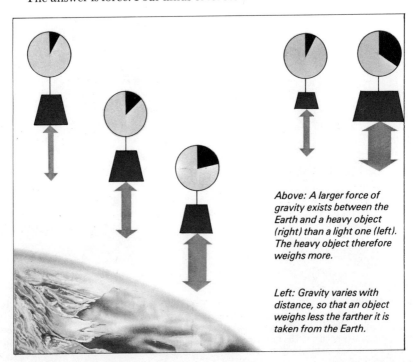

Above: A larger force of gravity exists between the Earth and a heavy object (right) than a light one (left). The heavy object therefore weighs more.

Left: Gravity varies with distance, so that an object weighs less the farther it is taken from the Earth.

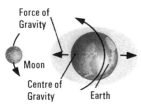

Above: Tides rise in the oceans because the Moon's gravity pulls up the water beneath the Moon. The Moon and Earth rotate around their centre of gravity, which is located inside the Earth. This rotation raises another tide on the side of the Earth opposite the Moon.

Above: The Sun's gravity affects the tides. When the Sun and Moon are in line, they act together to raise the big tides known as spring tides.

Below: When the Sun and Moon are at right angles, the Sun lessens the Moon's pull and the small tides or neap tides result.

necessary to explain the nature of matter from the tiniest part of the nucleus up to the structure of the Universe itself. The strongest force is that holding the nucleus together, and it is called the *strong interaction* because it is a strong force produced by the interaction of mesons in the nucleus. The strong interaction is 100 times stronger than the *electromagnetic interaction*, which is the force between electrically charged bodies (between the electron and the nucleus, or between atoms, or between molecules) or between two magnetized bodies. The third force holds together the particles in the nucleus that break apart when freed from the nucleus, producing radioactivity. It is therefore not very strong and is in fact known as the *weak interaction*, being a million million times weaker than the strong interaction. The final force is the force of *gravity* that pulls us to the ground, keeps the Moon orbiting the Earth and the Earth moving around the Sun, and holds together the solar system and the Galaxy. Because it acts to control the movements of such massive bodies, gravity might seem to be a very strong force. In fact, it is a very, very weak one and equal to only 10^{-40} or $1/10,000,000,000,000,000,000,000,000,000,000,000,000,000$ of the strong interaction!

The relative strength of these universal forces can, apart from the weak interaction, easily be demonstrated. Two pieces of metal small enough to be held in the hand exert no apparent gravitational force—it exists but is so weak that one does not apparently attract the other. However, if the pieces are magnetized, they will cling together. The electromagnetic force involved is obviously weaker than the strong interaction holding the magnets' nuclei together, otherwise the magnetic force would pull the nuclei apart and the magnets would disintegrate.

Energy

Energy is a word of which we hear a great deal nowadays; advertisements for certain foods assure us that they are brimful of energy and yet there is said to be an energy crisis. The word energy is used in so many different ways that it might appear to have several meanings. Although there are in fact several different forms of energy, energy itself can be simply defined in a few words: basically, it is the capacity to do work.

Anything that is doing work—for example, moving another object, heating it or pushing electric current through it—is using up energy in doing so. Therefore anything that is ready to do work—an engine about to pull a train, a pile of sticks ready for lighting, or a battery fully charged with electricity—possesses energy. When work has been done, it will have less energy. The object on which work is done—the train that begins to move, a poker placed in the fire that begins to glow, and a light bulb connected to the battery that lights—all these objects gain energy.

The Forms of Energy

Although one object loses energy as another gains it, energy is not created nor destroyed. The total amount of energy remains the same; it simply changes in form. There are many different kinds of energy, but they can be grouped into a few basic forms.

Potential energy is energy that an object has because of its position. A raised hammer, a wound-up spring and a drawn archer's bow all possess potential energy. This energy is ready to be changed into other forms of energy

A space rocket possesses an immense amount of energy that is ready for use as it stands on the launching pad (top). This energy is in the form of the chemical energy of the fuel inside the rocket. When the fuel burns, the chemical energy is turned into heat, a form of kinetic energy. The heat produces hot exhaust gases that drive the rocket upwards

Left: A nuclear reactor produces heat by changing the atoms in its fuel, turning uranium or plutonium into other elements. Engines that utilize chemical energy change the molecules in their fuel and the elements remain unchanged.

Below: An old steam locomotive changes chemical energy into kinetic energy. The burning of wood or coal in the boiler is a chemical reaction that produces heat, raising steam to power the locomotive.

and do work. When the hammer falls, it will drive in a nail; the spring, when released, will drive the hands of a clock and the bow will fire an arrow. As soon as any movement occurs, the potential energy decreases as it

Energy

changes into energy of motion (*kinetic energy*). Raising the hammer, winding up the spring and drawing the bow again uses up kinetic energy and produces a gain in potential energy. In general, the higher and heavier an object is, the more potential energy it has.

Kinetic energy is possessed by anything that is moving, and the faster an object moves, the more kinetic energy it has. Also, the heavier an object is, the greater its kinetic energy (but only when it is moving). Mechanical machines—cars, lathes, pile-drivers or any other machines that use a motor or an engine—produce kinetic energy, and this kind of energy is often called mechanical energy.

Electrical energy is energy that is given to an object by passing an electric current though it or giving it an electric charge. It is converted into mechanical energy in an electric motor or heat energy in an electric fire.

Chemical energy is energy that lies in the arrangements of atoms within molecules; by rearranging the atoms, chemical reactions occur and energy may be produced or taken up. Chemical reactions usually produce heat; a burning fire is an example. Chemical energy may also be changed to electricity in a battery and to kinetic energy in muscles.

Radiant energy is energy that can cross space. It includes light, radio waves and heat rays. Radiant heat is not the same as the kind of kinetic energy called thermal energy, but when heat rays strike an object they cause its molecules to move faster and it then gains thermal energy. Light and heat rays are produced by making objects so hot that they glow, as in the filament of a light bulb.

Nuclear energy is energy produced by changing atoms within a substance; it mostly appears as heat, either under control in a

ENERGY, WORK AND POWER

Energy and work are basically similar, work being the expenditure of energy. They are therefore both measured in the same units, the SI unit being the joule, which is named after James Prescott Joule (1818-89), the British physicist who helped to show that different forms of energy can be converted into each other without loss or gain of energy. Power is the rate of doing work or expending energy. An energy source which can do more work in a given time — make anything move faster or heat it more rapidly, for example — is more powerful than another energy source. The SI unit of power is the watt, named after James Watt (1736-1819), the British engineer who invented the first practical steam engine. One watt of power is produced as one joule of work is done in one second; a thousand watt (1 kilowatt) electric fire therefore uses up a thousand joules of energy every second.

Energy

Below: All the energy that is used on Earth comes from the Sun or from the atom. Every time it is used, energy is converted from one form to another, but the conversions eventually lead to radiation — mostly heat and light — that is radiated away into space

nuclear reactor or in an explosion in a nuclear weapon. The Sun produces its heat and light by nuclear reactions. Hence all life on Earth depends on nuclear energy and yet, in the threat of nuclear weapons, is endangered by nuclear energy too.

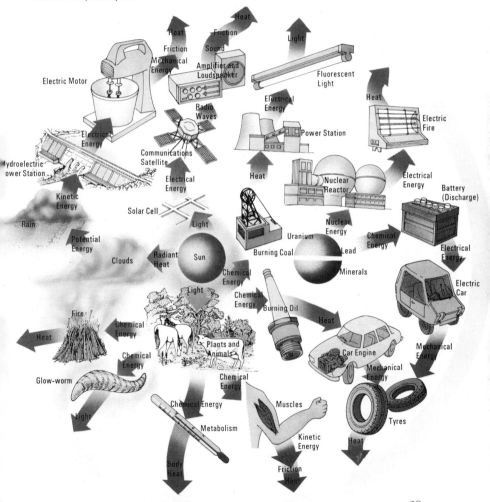

Force and Motion

Why does an arrow fired from a bow continue to move through the air after it has left the bow? This question perplexed people long ago, and it is easy to see why. At that time, things were made to move only if a man or a horse or ox pushed or pulled them—yet the arrow flew through the air without any aid at all. Another fundamental question about motion could also have occurred to them: why do things fall to the ground? Eventually one man answered both questions and although they may both seem simple, it took one of the greatest scientists the world has seen to solve them. This man was Sir Isaac Newton, the British physicist and mathematician who lived from 1642 to 1727.

Once set in motion, any object continues to move unless it is acted upon by a force. In a car, the occupants will carry on moving if the car stops suddenly. Unless they are securely fixed to their seats, they will fly forward and strike the windscreen if the car should crash. Here a car crash is simulated with dummy occupants to find out what dangers the driver and passengers would face in a real crash.

Newton discovered the three basic laws of motion that now bear his name. The first law answers the question about the flight of an arrow and states that **an object will remain at rest or will continue to move at the same speed in a straight line unless it is acted upon by a force**. That is, once an object such as the arrow has begun to move, it will keep on moving constantly until something either stops it or changes its motion in some way. No action of any kind is needed to keep it moving. This is why the planets always continue to move through space. They began to move when the solar system was formed, and it is not that there must be something in space to keep them moving but simply that there is nothing in space to stop them moving.

A force is therefore a kind of push that acts to begin motion or change motion. When a force acts, it uses energy and produces work. There are several kinds of force, just as there are several forms of energy. The expansion of anything when it is heated produces a force as its size increases; the power of the motor car, aircraft and rocket comes from the force of expansion of hot gases. The force produced

In a collision, one object imparts a velocity to another as momentum (mass x velocity) is transferred from one to the other. Momentum may also be carried through a line of objects in contact with each other. Here, the two red balls at each end of the line of five swing alternately to and fro. As one strikes the end of the line, its momentum is transferred down the line to the final ball, which is knocked away as if the first ball had struck it directly.

Force and Motion

by muscles comes from chemical changes in the muscles that make the muscle fibres contract. Electrical fields and magnetic fields both produce force.

Gravity is another kind of force that Newton himself discovered, simply by asking himself the second of our basic questions on motion—why do things fall to the ground? Galileo had found that objects accelerate as they fall; that is, their motion is changing. Newton realized that a force must therefore be acting on a falling body and this force is the pull of gravity. He said the answer came to him when he saw an apple fall from a tree, and the SI unit of force, which is called the *newton* after him, is by coincidence approximately equal to the weight of an apple. The pull of gravity exerts a force on everything at the surface of the Earth, and this force is the weight of anything.

MASS, WEIGHT AND INERTIA

The Apollo astronauts were weightless on their journeys to the Moon and had only a sixth of their normal weight while they stayed on the Moon. However, at no time did their *mass* change.

At a given point, the ratio of the masses of two bodies is directly proportional to the ratio of their weights (i.e. to the forces which they exert on whatever is supporting them). The actual weight of a body, however, depends not only on its position in space but also on its motion. Thus a man in space may have no weight because of his motion, although gravity still pulls him towards the Earth. A man on the moon will only weigh one-sixth of his weight on Earth, due to the smaller pull of the moon.

Inertia is the tendency of anything to resist a change in its motion and it is related to mass. Inertia has to be overcome to get an object moving and, once moving, it has to be overcome again to stop it moving. A massive object has great inertia, as you will know if you have ever pushed a motor-car. The astronauts were able to lope about on the Moon because their weight was low—that is, gravity did not pull them to the ground as strongly as on Earth. But inertia did not change because mass did not change. If the astronauts' moon-car had broken down, they would have had little trouble in lifting it as its weight would have been low. But they would have found that just as much effort would have been needed to push it back to the lunar module as would have been required on Earth.

Force and Motion

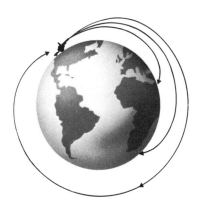

Friction

Another important kind of force acts only when objects are in motion, and it always acts to slow their motion. This force is called *friction*, and it is caused by contact with the medium in which an object is moving or with the surface on which it is moving. The friction produced in a medium is its viscosity.

Friction with a surface depends on the pressure between the object and the surface; the greater this pressure, the greater the friction. This can be easily demonstrated by sliding your finger along a smooth tabletop; if you press hard, friction increases and your finger comes to a stop.

Although it opposes motion, all forms of transport that run on wheels could not move without friction. Friction between the wheel and the ground enables the wheel to grip the ground and produce movement.

It might seem odd to say that no force is needed to keep something moving when a jet aircraft flies at a constant speed using its powerful engines. The reason is that the force of the engines pushing the aircraft forward is equalled by the friction with the

Above: Imagine a shell being fired from a more and more powerful gun (left). It would fall farther and farther away until, in theory, it would fall around the Earth and return to the gun. In fact, friction with the air would cause the shell to burn up. However, this is basically how a spacecraft orbits the Earth (right). It is launched into space with sufficient speed to send it on a path around the world. No air exists in space to slow the spacecraft, and once in orbit, it continues to travel around the Earth.

33

Force and Motion

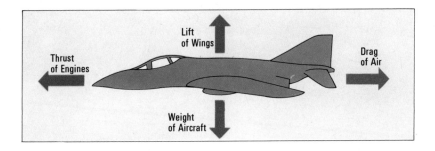

Above: Four main forces act on an aircraft as it flies through the air. The aircraft flies at constant speed if the thrust equals the drag, and at the same level if lift equals weight.

air through which the plane is moving; the two forces balance each other so that no overall force is acting on the aircraft and it therefore continues to move at a constant speed. Increase the power of the engines and the aircraft will move faster until friction increases to match the increased force; it will then be moving at a higher but constant speed

Momentum

An overall force acting on an object therefore increases or decreases its velocity. The resulting acceleration or retardation will depend on the size of the force and on the mass of the object. A greater force will produce a more rapid change in velocity but a greater mass will undergo a less rapid change. This is expressed in Newton's second law of motion, which can be stated as force is equal to mass multiplied by acceleration (or retardation). However, Newton preferred to use the quantity known as *momentum*, which is equal to the velocity of a moving object multiplied by its mass. The second law then states that **the rate of change of momentum is equal to the applied force**.

You are most likely to be reading this while sitting on a chair. The Earth's gravity is producing a force on you that is acting to pull you down. Yet you do not move. Another

Force and Motion

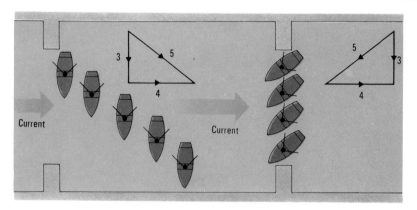

Above left: A boat heads across a river at 3 km per hour but is swept to one side by the current flowing at 4 km per hour. Its final velocity can be found by constructing a vector diagram in the form of a triangle in which two sides represent the speed and direction of the boat and current and the third side represents the final velocity, which is in a diagonal direction at 5 km per hour. Above right: The boat can move straight across if it heads into the current at 5 km per hour. The vector diagram shows that it will then be carried straight across the river at 3 km per hour.

force must be acting to oppose gravity, otherwise you would move. This force is an upward force in the legs of the chair, and it is called the reaction to the action of gravity. Newton's third law of motion deals with this situation and states that **action and reaction are equal and opposite**. Whenever any force acts, an equal force also acts in the opposite direction. Note that action and reaction occur only when one object exerts a force on another.

Adding Forces

Newton's law of motion and the equations of motion show how an object moves in a straight line; we have considered velocity and acceleration but not *direction*. When one force acts on a body, it will move in the direction taken by the force. But if two or more forces act at the same time, they may each act to

35

Force and Motion

push the object in different directions and the direction it finally takes will depend on the size of the forces involved and the direction in which each one acts.

If you try to row a boat across a fast-flowing river, you soon find that you cannot steer directly towards the point at which you wish to land because the current sweeps you downstream. You have to face upstream at a certain angle so that part of your motion opposes the flow of the current while the other part takes you across the river. Aircraft navigators have to allow for winds blowing them off course in exactly the same way.

Force and Motion

If two forces are added together but at least one of them is producing an acceleration, then the resulting motion will be a curve. This can easily be seen by throwing a stone into the air. The motion can be resolved or separated into two components at right angles, one horizontal and one vertical. The horizontal component keeps the stone moving forward at a constant rate (neglecting friction with the air), but the vertical component is affected by gravity. In a vertical direction, the stone rises and slows to a stop and then falls with increasing speed until it hits the ground. Combining these components, the stone's final motion is a curve called a *parabola*, and it hits the ground some distance away. Its range depends on the angle at which it is thrown as well as its speed. Archers and gunners can make precise estimates of the paths of projectiles—in their case, arrows or shells.

Any object that is moving in a circle—a satellite, a car going round a corner, a stone being whirled around on a piece of string—is constantly changing direction and is constantly being accelerated in any one particular direction; it is therefore always subject to a force. The force is directed towards the centre of the circle and it is called *centripetal force*. In the case of a satellite, the centripetal force is the Earth's gravity pulling it into an orbit; with the car, it is the friction of the tyres with the road steering it into a circle, and with the stone, the centripetal force is the tension in the string. All act to prevent the object from leaving its circular motion.

Rotation

An object spins when it is given a particular combination of forces called a *couple*, which consists of a pair of forces in opposite directions

At any instant, a rider and chairoplane seat, considered as one body, are subjected to two external forces. One is the force of gravity pulling them towards the centre of the earth and the other is the tension in the supporting cable. This latter force can be divided into two components, one acting vertically upwards and the other acting horizontally towards the axis about which the rider and seat are rotating. The vertical component balances the downward pull of gravity so that the rider remains suspended in the air, remaining in the same position relative to the support. The horizontal component provides the force necessary to produce circular motion.

Force and Motion

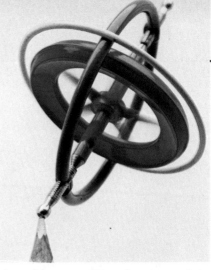

Above: A gyroscope circles around its pivot because of precession.

applied to opposite sides of the object. Spinning a coin demonstrates how a couple is applied. Rotating objects possess some unusual properties, for example a kind of momentum called *angular momentum*. This is related to the mass, size and spin of the rotating object: the greater its mass, the larger its diameter, and the faster its speed of rotation, the greater its angular momentum. Angular momentum remains constant in a spinning body if no force acts to speed it up or slow it down. This means that if a rotating body changes its diameter, then its speed of rotation must also change so as to conserve angular momentum. This can be seen when spinning ice-skaters suddenly draw in their arms. They immediately spin faster because they are reducing their diameter.

The gyroscope is a particular kind of rotating body. It is simply a spinning disc mounted in a frame, rather like a bicycle wheel. Conservation of angular momentum ensures that the axle of the gyroscope does not change its direction once it has been set spinning. The gyroscope can therefore be used as a compass. A gyrocompass indicates the direction in which a ship or aircraft is heading much more accurately than a mag-

Force and Motion

Right: This toy figure always springs up right no matter how much it is tilted. The centre of gravity is so low that it always remains to one side of the point on which the toy pivots and gravity pulls it back upright.

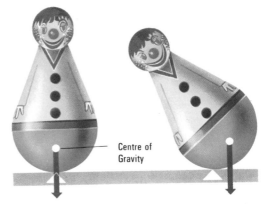

Centre of Gravity

netic compass, and it also acts as a controller for automatic navigation devices.

Another characteristic of spinning bodies, and one particularly exhibited by the gyroscope, is precession. If the axle of the gyroscope is forcibly turned, then it will move in a direction at right angles to the direction in which the turning force is applied. If a toy gyroscope is set spinning and one end is placed on a pivot, then the gyroscope will begin to keel over as the force of gravity acts vertically on the axle.

THE PENDULUM

A pendulum swings to and fro at a constant rate, every swing taking exactly the same time, provided the swings are not too large. It would continue to swing for ever if friction did not eventually bring it to a halt. The pendulum is therefore a good device for regulating clocks and has been used for this purpose for three centuries. The period of the swing of a pendulum depends on its length but not its mass, and the time-keeping of a pendulum clock is adjusted simply by raising or lowering the weight at the end. The period also depends on the force of gravity, and a pendulum can be used to make an accurate determination of gravity.

Relativity

Sometimes you can sit in a train at a station and think it has started to move, and then suddenly realize that you are still stationary and another train alongside is leaving in the opposite direction. For a moment, you are not sure whether you are moving or not; all you know is that one train is moving relative to the other. All motion is in fact relative; there is no central stationary point in the Universe against which motion can be measured. Your train may be travelling over the ground, but the ground itself is moving as the Earth circles the Sun. Furthermore, the Sun is moving as the Galaxy spins and the Galaxy itself is also in motion. Everything in the Universe is moving, and consequently motion cannot be measured absolutely; all one can say is that one object is moving at a certain velocity relative to another object.

From this idea, and from the fact that the speed of light is always observed to be the same no matter how fast the observer is moving, Albert Einstein deduced some astonishing conclusions. These were presented in the *Special Theory of Relativity*, which was published in 1905 and revolutionized physics. The main conclusion was that length, mass and time are all affected by motion. If an observer looks at an object or system of objects in relative motion to him, he will observe that length decreases in the direction of motion, mass increases and time slows down the faster the system moves. The effects are too small to be noticed at the speeds we observe on Earth, and Newton's laws of motion still hold

Two spacecraft pass one another, each travelling at 30 km per second as measured from Earth. But each spacecraft will be approaching the other one at 60 km per second. Their velocities are 30 km per second relative to Earth, but 60 km per second relative to each other. However, if observers on both spacecraft and on Earth then measured the velocity of light coming from the Sun, they would all get exactly the same figure — nearly 300,000 km per second. This illustrates the two postulates of the Special Theory of Relativity: that all motion is relative and that the velocity of light is always constant, no matter how it is measured.

good. But if the relative motion reaches the speed of light (300,000 kilometres a second), then in the observed system, length becomes zero, mass infinite and time slows to a stop. Clearly, such a situation is impossible and so Einstein concluded that relative motion at or beyond the speed of light is impossible; that is, nothing can move faster than light. Time slowing and mass increase have since been observed in atomic particles that move at very high speeds.

If an atomic particle is given more energy, it travels faster. As more and more energy is pumped in, it will approach the speed of light. Its mass then increases substantially but its speed does not get much greater. The extra energy mainly produces an increase of mass. Einstein deduced that mass and energy are related to one another by the famous equation $E = mc^2$, in which E is energy, m is mass and c^2 is the speed of light multiplied by itself. As c^2 is such a huge quantity, a little mass is equivalent to a vast amount of energy. This was later shown to be true when nuclear power was released.

The Special Theory considers systems in relative motion at a constant velocity. Einstein went on to consider the effects of acceleration, and published his conclusions as the *General Theory of Relativity* in 1916. The basic ideas behind this theory are that gravity is equivalent to acceleration, and that light bends in an accelerating system.

This conclusion can be thought of in a more general way: that gravity distorts space. Light or anything else travelling through space finds its path curved by a gravitational field because distance is shortened by gravity. The effects are noticeable only in a strong field of gravity such as that of the Sun.

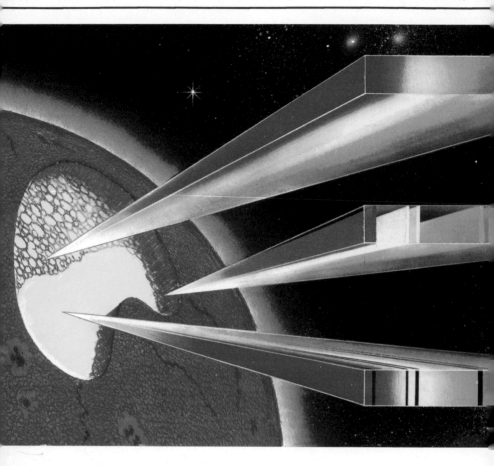

Light from the visible surface of the Sun originates as a continuous spectrum of colours (top) that combine to give white light. However, close examination of the Sun's spectrum reveals the presence of dark lines (bottom). This happens because certain elements absorb certain wavelengths as light passes through the Sun's atmosphere. This kind of spectrum is called an absorption spectrum. Each element produces a set pattern of lines and the absorption spectrum therefore identifies the elements present in the Sun's atmosphere. The elements in the atmosphere also emit light at the same wavelengths, giving an emission spectrum (centre).

The Electromagnetic Spectrum

If you place a glass prism in sunlight, you will find that a band of coloured light is produced. The white sunlight is split up into a spectrum of several colours. Like the colours of the rainbow, which is a natural spectrum, the colours range from red through orange, yellow, green and blue to violet. If a thermometer is placed just beyond the red end of the spectrum, it will show a slight rise in temperature. This is because the Sun's light contains invisible heat rays that lie beyond red in the spectrum and are consequently known as infra-red rays. Other kinds of detectors will show that there are more invisible rays beyond the infra-red; first come microwaves and then radio waves. Other invisible rays lie beyond the violet end of the spectrum—ultra-violet rays, X-rays, gamma rays and cosmic rays. In fact, visible light makes up only a very small section of this great spectrum of rays.

These rays have many different uses and may seem to be entirely different. But they all belong to one great family of rays called electromagnetic radiation. This radiation is composed of vibrating electric and magnetic fields, and it travels in a wave motion through empty space and through transparent sub-

The Electromagnetic Spectrum

stances. For example, glass and water are transparent to light just as our flesh is transparent to X-rays and the walls of our homes are transparent to radio waves. The whole electromagnetic spectrum is made up of energy that moves in a wave motion at a fixed velocity (that of light). The only difference between the various kinds of rays is their particular wavelength—the distance between each maximum of energy as the wave passes. The range of wavelengths is enormous —from less than a hundred-million-millionth of a metre for cosmic rays to thousands of metres for radio waves. Light occupies the very small wavelength band of 4 to 7 ten-millionths of a metre. The various rays may also be identified by their frequency—their rate of vibration—as well as by wavelength. Electromagnetic frequencies vary from ten thousand vibrations a second for radio waves up to a million million million million vibrations a second for cosmic rays.

Radar – RAdio Detection And Ranging – makes use of micro radio waves to 'observe' distant objects. When a beam of transmitted waves hits an object, the waves bounce back to the receiver where the information appears as dots of light or 'blips' on a screen.

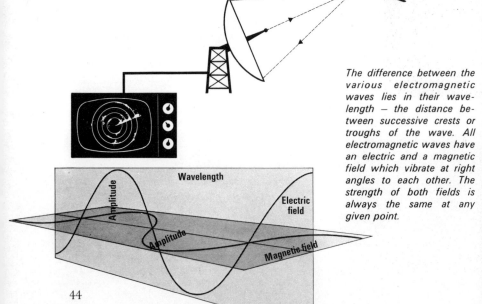

The difference between the various electromagnetic waves lies in their wavelength – the distance between successive crests or troughs of the wave. All electromagnetic waves have an electric and a magnetic field which vibrate at right angles to each other. The strength of both fields is always the same at any given point.

The Electromagnetic Spectrum

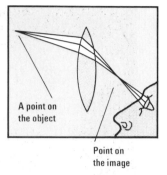
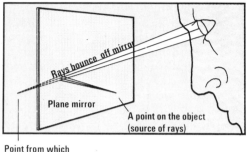

Left: A converging lens gives a real image of an object placed some way away from it. Rays of light coming from a point on the object cross after passing through the lens; the eye, which receives rays spreading out from the crossing point, sees an image at that point. Right: A plane mirror deflects the rays from an object from their direct path. It produces the illusion of an image as far behind the mirror as the object is in front.

Sir Isaac Newton

Properties of light

Scientists once believed that light was made of minute particles called corpuscles. But the Dutchman Christiaan Huygens put forward the theory in the 17th century that light is made up of waves. Today scientists think that light, like other forms of radiation, is made of quanta – tiny 'packets' of energy which travel in waves. Light travels normally in a straight line. But it can be deflected or bent from its straight course by reflection and by refraction.

Light is reflected by shiny surfaces such as mirrors. The most common is the plane (flat) mirror. The image you see in a plane mirror is always what is called a *virtual* one – it is imaginary and unreal. The rays are obeying the laws of reflection by which the angle of incidence equals the angle of reflection. The angle of incidence is the angle between the rays hitting the mirror and the normal – a line at

45

The Electromagnetic Spectrum

right angles to the mirror. The angle of reflection is the angle between the rays bouncing off the mirror and the normal. Rays bouncing off a plane mirror appear to be spreading out from a point as far behind the mirror as the object is in front. There is in fact nothing at the point from which the rays seem to be coming. The image is therefore unreal or virtual (see diagram on previous page).

Curved mirrors are also widely used. Again the rays must obey the laws of reflection. But the distance from the image to the curved mirror is not equal to the distance from the mirror to the object. Curved mirrors produce images that are larger or smaller than the object. There are two main types of curved mirrors: concave, which curve inwards like a saucer, and convex which bulge out. Both kinds are widely used. Concave mirrors magnify for shaving and for astronomical reflecting telescopes. Convex mirrors reduce for car wing mirrors. You can also see the amusing and rather alarming effects curved mirrors can produce in a Hall of Mirrors at a funfair.

Changing the medium

Light can also be bent or deflected by refraction. Light is refracted when it passes from one medium to another – for example, from air into water, or from air into glass. Refraction occurs because light has a slightly different speed in every medium. When the speed changes, the light ray bends. The denser the medium, the more the light is bent. A pencil half in, half out of a glass of water appears to be broken as it enters the water. This is because the light slows down at the water's surface.

The phenomenon of refraction is made use of in lenses and prisms. Lenses are specially

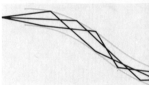

Fibre Optics depends on the princi that a light ray striking a surface a large angle of incidence will be reflected from the surface. In this way light rays will travel along gla or plastic fibres, being continually reflected from one wall of the fibr to another (above). A group of fib will carry light from one end to the other, producing a spray of light a the fibre tips (left).

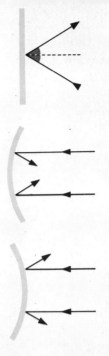

Reflection occurs because light rays bounce from surfaces. A plane mirror (above) produces an image the same size as the object. In all mirrors, rays are reflected so that the angle of incidence equals the angle of reflection (top right). Concave mirrors make light rays converge, producing a magnified image of a near object (centre right), and convex mirrors cause rays to diverge, giving a diminished image (bottom right).

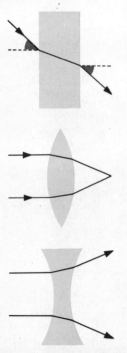

Refraction occurs where light rays pass through surfaces. On entering a denser medium, rays are bent towards the normal (top left). As they leave, the rays bend back again, leaving at the same angle as they entered. By bending light rays, refraction causes images to be displaced, as when objects are seen through water (above). Lenses work by refraction, causing light rays to converge or diverge, producing magnified or diminished images accordingly.

The Electromagnetic Spectrum

curved pieces of glass. A concave or diverging lens is curved inwards – it is thinnest at the middle. A convex or converging lens (such as a magnifying glass) is curved the other way. A converging lens will give rise to an unreal or virtual image of any object placed *near* it. The image appears to be on the same side of the lens as the object but at a point behind the object. A converging lens will produce a real image of an object placed a long way from it. Light rays pass through the lens and cross and we see rays of light spreading out from the crossover point. At this point we also see the image – on the opposite side of the lens from the object. A concave or diverging lens always produces a diminished and unreal image of an object no matter where the object is situated. If the rays are traced back they come from a point in front of the object.

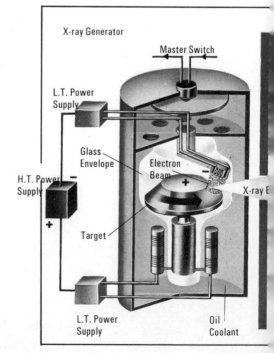

An X-ray machine in use in a hospital. The X-ray beam passes through the patient's chest and strikes the screen of an image intensifier. The screen produces an electron beam that causes a picture to form on a fluorescent screen viewed by the doctor. The intensifier enables low doses of X-rays to be given.

X-rays

In the electromagnetic spectrum, waves with the smallest wavelength have the highest energy. Beyond the visible spectrum, beyond even the ultra-violet, is the region of penetrating, high-energy radiation, both manmade and natural. At 3×10^{-10} metres, X-rays are first encountered. By 3×10^{-12} metres, the X-rays are so 'hard', or penetrating, that they are designated γ-rays and become too dangerous for medical use.

X-rays were first discovered by Röntgen in 1895. Within months they were being used to show up some lead pellets accidentally shot into a New York lawyer's hand. The principle of the medical X-ray photograph is as simple as a shadowgraph. Although X-rays pass through all human tissue, they are at least partially absorbed by the solid matter of the bones.

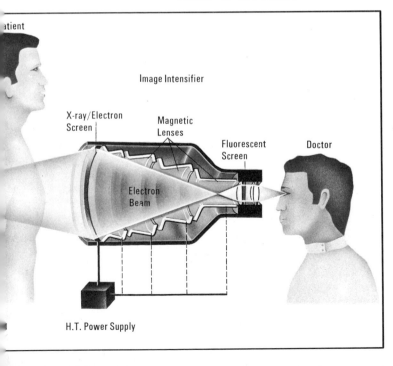

Lasers

The light produced by the laser is clearly no ordinary kind of light. Lasers can give an intense beam that will cut out patterns in steel plate and drill holes through diamond, the hardest substance known, to make dies for drawing wire. Military men have their eye on the laser to produce a kind of death ray. But so do medical men; laser beams can be used to drill teeth painlessly, and to repair damage to the retina of the eye.

The name laser stands for Light Amplification by Stimulated Emission of Radiation; this mouthful does in fact describe how the laser works, but not why its light is so special. The laser can perform its many miracles because it produces a beam of coherent light. In this kind of light, all the waves vibrate exactly together. All the waves reinforce each

Right: Sparks fly as a laser beam cuts its way through a steel sheet. Below: The crystal at the heart of a laser produces an intense beam of light that is reflected between mirrors until it has sufficient intensity to leave through the half-silvered mirror.

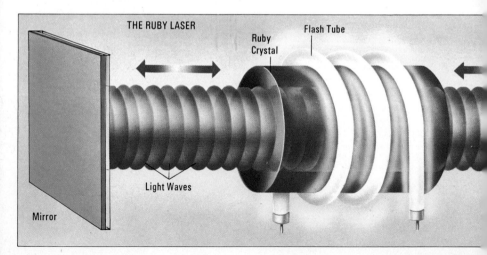

THE RUBY LASER
Ruby Crystal
Flash Tube
Light Waves
Mirror

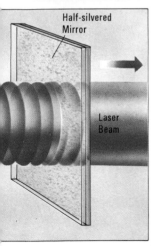

other, producing light of great energy. The light is of a single wavelength, but each kind of laser produces its own wavelength. Some lasers produce invisible radiation, such as infra-red rays.

Light is produced when the electrons in an atom jump from one orbit around the nucleus to another nearer the nucleus. In doing so, they lose energy and this energy is emitted from the atom as light. To get up to the higher orbits, the electrons must be given energy. Now, if all the atoms in a substance can be made to produce light at the same time, then a pulse of coherent light will be produced. This happens in the laser.

Lasers

At the heart of the laser is a crystal or tube of gas into which energy is pumped by surrounding it with a flashing light or a source of radio waves or electrons. More and more electrons in the atoms are raised to high-energy orbits by this action. Suddenly, an atom returns to its low-energy state and gives out light. This starts off a cascade of light production from all the other high-energy atoms. An intense pulse of light builds up and leaves one end of the laser. The light is amplified in this way, and it is done by stimulating the emission of radiation (light) from high-energy atoms with the light produced by the initial atom. This is why the laser is so named.

The beam produced is straight and very narrow, and can be used for precise alignment of tunnels and pipelines. It hardly spreads at all, reaching a width of only three kilometres at the distance of the Moon. Laser beams are fired at mirrors left by the Apollo astronauts to give a very accurate measure of the Moon's distance. A laser beam can also be modulated to carry information, rather as a radio wave is modulated (see page 60). Because of the higher frequency of laser light, it can carry vast amounts of information. On Earth, fog and mist would block the beam but for communications in space, the laser would be ideal.

A laser interferometer is an instrument for measuring minute distances by interference between two laser beams (see page 210). It can detect minute movements, such as those across geological faults and in dam walls. A laser gyroscope that has no moving parts but can detect changes of direction works in a similar way.

Nuclear energy may also benefit from the laser. Thermonoculear fusion could possibly be triggered off by laser beams.

Below: A laser crystal or tube contains atoms (1) that are excited, or raised in energy, by bombarding them with light or other radiation (2). An excited atom then suddenly loses its extra energy by producing light. This light strikes other excited atoms, causing them to lose their extra energy in the same way (3). The light produced is reflected between mirrors (4), causing a build-up of light as all the excited atoms lose their energy. Finally the light leaves through a half-silvered mirror at one end of the laser (5).

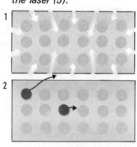
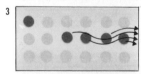
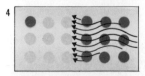
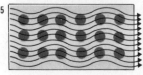

Lasers

HOLOGRAPHY

The most amazing effect produced by lasers is *holography,* by which solid images can be reproduced. Because laser light is coherent, it contains information on the shape of an object as well as its appearance when it is reflected from an object. This reflected light can be captured on a photographic plate to form a hologram. In ordinary light, it looks like a pattern of whorls. But when the hologram is illuminated by laser light, an image of the object is reconstructed with its shape. It not only appears solid but *is* solid; you can walk around the image and its perspective will change!

Holographic images can be made in colour, and it may one day be possible to have holographic television. Such a system could produce a totally life-like image that would occupy a corner of the room just as if it were real.

Below right: A 1000-joule laser chain being used for research into fusion power. Inside each amplifying laser is an oval disc made of special glass containing atoms of the element neodymium. Around this are powerful flash lamps which provide the excitation energy for amplifying the laser beam.

Holography has many practical uses. Below is an image of a car tyre made by superimposing a holographic picture of the unpressurized tyre on another holographic picture of the tyre under pressure. The differences between the two pictures show up as interference fringes (light and dark bands) that mark the distortions on the pressurized tyre. A bad distortion can be seen in the 5 o'clock position.

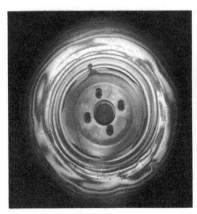

53

The Spread of Colour

We often use the phrase 'all the colours of the rainbow' in describing something that is multi-coloured. There are indeed many colours in the rainbow, or the spectrum of sunlight, and each corresponds to a certain wavelength of light. Red has the longest wavelength and violet the shortest, and other colours range between them. But there are many colours that are not present in the rainbow—pink and brown for example. Where do they come from?

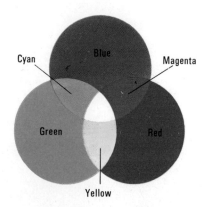

Additive mixing of colours occurs in coloured light, as in theatre lights and colour television. White forms by a combination of all three primary colours, and three secondary colours form by combining two primaries.

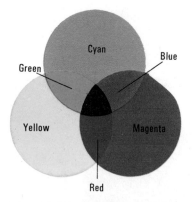

Subtractive mixing occurs in paints, inks and dyes. The three basic colours are the three secondaries of additive mixing. Combinations of two basic colours produce the three primaries, and all three basic colours combine to give black.

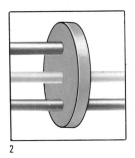 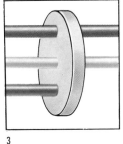 

2 3 4

Colour filters produce colour by subtractive mixing in the same ways as paints and inks. White light is a mixture of red, green and blue. A magenta filter absorbs green and passes red and blue, which combine to give magenta (1). Similarly a cyan filter absorbs red and passes green and blue that combine to give cyan (2), and a yellow filter absorbs blue and passes red and green that combine to give yellow (3). All three filters together absorb all the colours in white light (4). The layers of dyes in colour photographs act as filters in the above ways.

An object appears to have a certain colour because its surface reflects light of that colour. White light, or sunlight, is a mixture of all colours, and a red object appears red in white light because its surface reflects red but absorbs all the other colours in the white light illuminating it. Similarly a white object would appear red in red light, because its surface always reflects all the light illuminating it. If we looked at it in blue light, it would appear blue. But if we looked at the red object in blue light, it would appear black because the surface would be reflecting no light. This is why everything appears so lacking in colour beneath the orange glare of sodium-vapour street lights. The sodium light is of a single colour—yellow-orange—and surfaces either reflect this colour or absorb it, so that everything appears yellow-orange or black.

Coloured lights work in the same way. If we place a red filter over a white light, then the filter will absorb all colours except red; only red light passes through the filter, producing a red beam.

Few surfaces or filters absorb all but a single wavelength of white light. Most reflect or pass a mixture of colours, and these colours combine in the eyes to produce a vast range of colours.

The Spread of Colour

It seems that the eye is mainly sensitive to three basic or primary colours—red, green and blue. Other colours are combinations of these three primary colours, and this is made use of in many forms of colour reproduction—television, for instance.

Mixing Colours

Additive mixing is done with sources of light—theatre lights or the glowing screen of a colour television set. If you look at a colour set close-to, you will see that the picture is made up of a pattern of tiny dots or thin stripes of colour and that only three colours are present—red, green and blue. From a distance, the dots or stripes merge together to form a picture and the primary colours merge to give full colour. If all three colours are equally bright, then white is seen, becoming grey as the brightness is reduced. Black is an absence of all colours. Three secondary colours are produced by combinations of two primary colours—yellow from green and red, cyan from green and blue, and magenta from red and blue. All other colours are combinations of all three primary colours. The actual shade will depend on the brightness of each primary colour.

Subtractive mixing occurs when paints, inks or dyes are used. It is obvious that additive mixing does not work for these, as green and red make brown when you mix paints and mixing green, red and blue gives a muddy brown, not white. An absence of colour gives white—as long as you are working on white paper. The colour mixing is different because the pigments and dyes absorb colour from the light illuminating them. Dyes or pigments of the secondary colours form the basic colours because they absorb single primary colours; yellow pigment absorbs only blue, leaving

Microphones pick up the sound of the actor's voice and the cameras record the picture before them, turning sound and vision into electric signals. The signals are amplified and then fed to the transmitter.

green and red to be reflected or passed, and to combine to give yellow. Cyan pigment absorbs red, leaving blue and green to combine; and magenta pigment absorbs green, leaving red and blue to combine. Red is obtained by mixing yellow and magenta, absorbing blue and green respectively and leaving only red to be reflected or passed. Mixing all colours equally results in the pigments absorbing all the colours in the illuminating light, and black is produced.

In colour photography, light passes through layers of yellow, cyan and magenta dyes in the film or surface of the print. In colour printing, a picture is made up of dots of yellow, cyan and magenta, which combine to give a full-colour picture at a distance. This method does not give a deep black, and so black is added as a fourth colour to produce a life-like result.

Radio Waves

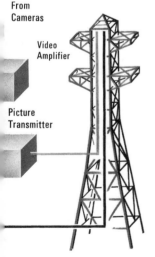

From Cameras

Video Amplifier

Picture Transmitter

Early in the nineteenth century, Michael Faraday showed that electric currents generate *magnetic* fields when they flow. Later, in 1864, James Clerk Maxwell proved mathematically that electrical disturbances produce effects *at a distance*. He showed that electromagnetic energy could move outwards in the form of waves, thereafter travelling at the speed of light. The way was thus opened for communication by radio.

Heinrich Hertz, in 1888, was the first to produce such an effect. He caused a spark to be triggered by radio emission from a larger more powerful spark discharge placed some small distance away. Guglielmo Marconi followed up these experiments in the firm belief

Radio Waves

Video Wave (Amplitude-modulated)

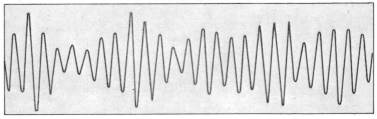

Audio Wave (Frequency-modulated)

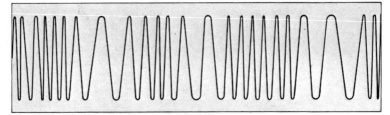

At the transmitter, the signals are combined with radio waves known as carrier waves. The signals vary depending on the patterns of sound and light being transmitted, and these variations are impressed on two carrier waves by varying the amplitude of one (top) in step with the video (vision) signal and the frequency of the other (bottom) in step with the audio (sound) signal.

that electromagnetic waves could be used to transmit messages. He devised an apparatus in which the spark gap was connected between an elevated wire (in effect, an aerial), and the Earth. His receiver was placed at first only nine metres away. But gradually he moved it, with continued success, two hundred and seventy-five metres away, and finally to three kilometres. By 1901 he had not only bridged the English Channel, but even sent the morse letter 's' from Cornwall, England to Newfoundland, Canada.

Nowadays, very complex radio signals are

Radio Waves

Above: The giant satellite-communications aerial at Goonhilly, Cornwall. Communications aerials pick up messages from man-made satellites in the same way as radio telescopes which 'listen' to waves transmitted from distant stars.

Radio Waves

sent by modifying the uniform waves produced by an alternating electric current. In effect, the transmitter sends out a uniform radio signal and the electronics are arranged so that speech or television pictures are impressed upon the radio waves. The process of coding the (carrier) radio wave is known as *modulation*, and involves either altering the amplitude of the wave, or its frequency. Speech itself is represented by an irregular wave pattern and the transmitter is designed so that the carrier wave varies in frequency and so duplicates the pattern of the broadcast speech or music. This is known as *frequency modulation*.

At the receiver, electrons in the aerial vibrate sympathetically with the received radio wave. Thus the flow of electricity which they comprise forms an alternating current. The signal is then *amplified* by electronic means. The coded speech signal, too, must also be untangled from its carrier and this is achieved by a *demodulator*. Further amplifiers increase the strength of the reconstituted signal, so that ultimately it is strong enough to drive the coil of a loudspeaker and be converted into sound.

Television Broadcasting

Television pictures, too, may be sent via a carrier wave, but for this *amplitude modulation* must be used. The scene in the television studio is recorded by the camera, in which the scene appears as a series of lines. On each line the amount of darkness and light varies. Many times a second, the image in the camera is scanned by an electron beam. Dark and light parts of each strip return differing numbers of electrons. These are received by the transmitter and then amplified. The next step is to imprint the corresponding pattern upon

Right: The carrier waves are received by the aerial at home. The aerial produces electric signals varying with the carrier waves, and these are separated into video and audio signals. They are then demodulated (the carrier waves are removed), leaving the original video and audio signals as produced by the cameras and microphones in the studio. These are fed to the tube and loudspeaker of the television set, which reproduce the picture and sound in the studio.

Radio Waves

the radio wave. For this, an ultra high frequency (UHF) carrier wave is used with a frequency in the range of 3×10^8 to 3×10^9 hertz (cycles per second).

The picture is encoded by varying the *amplitude* of the wave. Between certain pulses, the waveform is made to vary so that when high crests are transmitted, a corresponding 'dark spot' is indicated. When low crests are transmitted a corresponding 'bright spot' is indicated.

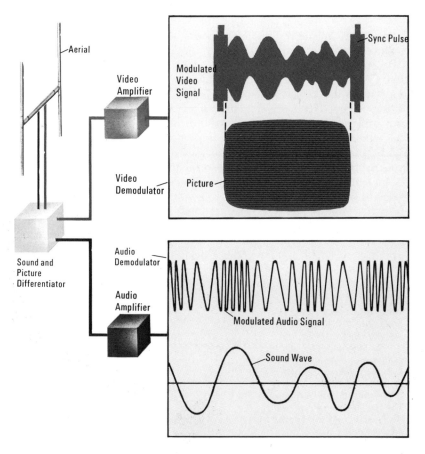

Radio Waves

In the camera, the scanning beam moves across a horizontal line of the screen. At the receiver, a similar electron beam traverses the domestic television screen. As this moves along its path, it excites small phosphorescent spots. If the electron beam is intense it records a bright spot in the scene.

The timing for the scanning beams in studio camera and television receiver is kept strictly synchronized. The pulses mentioned earlier act as coded time signals. In terms of the waveform, these 'sync pulses' appear as regular square-topped peaks sent out by the transmitter. Whenever one is received, the scanning beam is instructed to start another strip on the domestic screen. When it returns to the beginning it is moved fractionally lower as well. In this way it covers the whole television screen. Normally 625 such lines are involved in generating a UHF picture and the whole of the scene is covered more than thirty times each second.

Colour television also works on this principle, but it can be thought of as three such systems working in parallel. The camera in a colour television studio registers the brightness of points within the scene in terms of the three primary colours—red, green and blue. In practice, so as to accommodate black-and-white as well as colour receivers, the three are added together and a single 'brightness' signal

Above: The planet Jupiter is an intense source of radio waves, produced by the strong magnetic field that surrounds the planet.

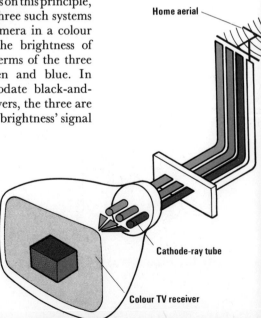

The television receiver *sorts out the three colour signals, and electron guns in the cathode tube shoot scanning beams on to the television screen. The screen is coated with three kinds of phosphor dots which glow respectively red, green, and blue when struck by the electron beam.*

Radio Waves

Transmitting in Colour
Separating mirrors split the light from the scene into the three primary colours – red, blue, and green. Three image orthicons change the light rays into electronic signals for each colour. Before transmission, colour signals are strengthened by the encoder while the adder forms black and white signals.

is transmitted as before. Nevertheless, mixed in and on the same carrier wave, there is information about the 'redness', 'greenness', and 'blueness' of each spot of the scene. At the receiver screen, three different phosphorescent substances are used which will reproduce all the colours.

The frequency of the carrier wave for the transmission of the picture is necessarily higher than that for the speech channel. The reason for this is that more information is involved.

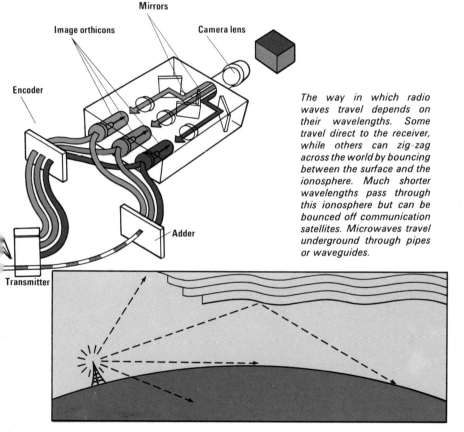

The way in which radio waves travel depends on their wavelengths. Some travel direct to the receiver, while others can zig-zag across the world by bouncing between the surface and the ionosphere. Much shorter wavelengths pass through this ionosphere but can be bounced off communication satellites. Microwaves travel underground through pipes or waveguides.

Electricity & Magnetism

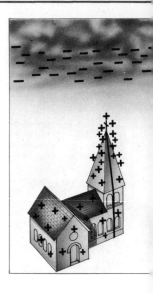

ELECTRICITY is the most useful form of energy there is. It is easy to produce; it can be transmitted over long distances; it is clean to use and has no smell. Above all it is convenient.

The electricity produced by nature lightning – is a different kind of electricity from that which flows through an electric-light bulb. It is called static electricity, because it exerts a force when it is stationary.

The other kind of electricity needs to flow to have any effect. The electricity in a battery will not make a light bulb glow until bulb and battery are linked by wires through which the electricity can flow. This kind of electricity is often called current electricity; the wire 'channel' through which it flows is known as a circuit.

Current electricity is produced by converting some other form of energy into electrical energy. The most common electricity pro-

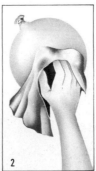
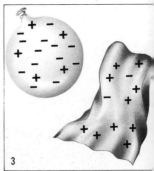

64

1. A lightning conductor consists of a pointed metal rod mounted on the roof of the building to be protected. The rod is connected by a wire to the ground. When a negatively charged cloud passes over the building, it repels the negative charges (electrons) in objects beneath it. The top of the building thus becomes deficient in electrons, and it acquires an overall positive charge. In any conductor, electric charges tend to pack tightly together at the most highly curved parts. For this reason, the pointed top of a lightning conductor acquires an extremely high positive charge. This charge ionizes the surrounding air — that is, it splits up the gas molecules of the air into charged particles. 2. The positively charged particles are then repelled from the point of the lightning conductor in a steady stream called an electric wind. The negatively charged cloud attracts these positive particles, which gradually cancel the cloud's charge, so that it eventually becomes electrically neutral. In most cases this prevents the building being struck by lightning. But, if the cloud has an extremely high charge, lightning may result. 3. If the building has a lightning conductor at its highest point, the lightning strikes this, and the charge flows safely down the wire to the ground. 4. When lightning strikes an unprotected building, the electric charge from the cloud passes through it, causing considerable damage.

uncharged balloon and cloth each ain equal numbers of positive and tive charges. 2. When the balloon is ed with the cloth, electrons are ferred from the cloth to the balloon. e balloon thus acquires an excess of rons, and therefore has an overall tive charge. The cloth, having lost rons, has an overall positive charge. cloth and balloon will therefore attract other. If two balloons are charged in vay, they will repel each other.

65

battery and the generator. In a ... al energy is changed into ... In a generator mechanical ...ged into electrical energy, the ...ical energy being supplied by some kind of engine or turbine.

Conductors

Most of the electrical components in an electric circuit are linked by copper wires. Apart from silver, copper is the best passer-on, or conductor of electricity. Most metals are fairly good conductors.

The electrons flow in a conductor when 'pressure' is applied to them, the source of electric pressure being a battery or generator. The term for electrical pressure is voltage. The extent electricity will flow in a conductor when a given voltage is applied depends on how that conductor resists the flow of electricity (its electrical resistance)

Liquids and gases may also conduct electricity. Then it is not electrons but ions (atoms which have gained or lost electrons) that move to carry the current. Gases do not normally conduct electricity. They have to be ionized first. When current flows through the gas, it gives rise to a light, or discharge. Neon lighting is a typical example of a gaseous discharge.

In contrast with conductors, there are materials known as insulators, or dielectrics which conduct electricity badly if at all. Rubber and plastics are good insulators. That is why they are used to cover electrical wiring. Glass and porcelain are also excellent insulators. They are used, for example, to isolate high-voltage transmission lines from the steel pylons which support them.

Some substances are neither conductors nor insulators. They allow a limited flow of electrons only. They are called *semiconductors*.

Above: The voltaic pile was invented by Alessandro Volta. He found that a layer of salt impregnated cloth, sandwiched between a piece of zinc and a piece of copper, produced an electric current. Several of these cells in a pile produced an even stronger current. The voltaic pile was the first battery.

Electricity and Magnetism

Germanium and silicon are the best known of these materials, which are used in electronics for making transistors and similar devices.

Electrochemical effects

Electrical and chemical energy are convertible – and in both ways. Not only can a chemical reaction produce electricity, as in a battery, but electricity can bring about a chemical change.

A simple battery can be produced by putting two dissimilar metals (termed electrodes) into a solution that conducts electricity (called electrolyte). The so-called simple cell consists of electrodes of zinc and copper in an electrolyte of dilute sulphuric acid. When connected in a circuit, current flows from the copper (positive) to the zinc (negative) electrode. The simple cell soon ceases to work because bubbles of hydrogen blanket the copper electrode. Modern batteries, such as the ordinary $1\frac{1}{2}$-volt dry cell which has a carbon positive and zinc negative electrode, contain substances which remove the hydrogen.

The simple cell and dry cell are called primary cells – once they are run down they are finished. Certain so-called secondary cells can be recharged by passing electric current through them from an outside source.

The car storage battery consists of six sets of secondary cells. The battery has alternate plates of lead (negative) and lead dioxide (positive) in an electrolyte of sulphuric acid. The battery discharges – electricity is produced – when the lead and lead dioxide react with the acid to form lead sulphate. Charging the battery with electric current changes the lead sulphate back again to lead and lead dioxide respectively. The battery in a car is constantly being charged by the generator.

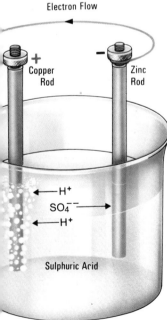

Below: A simple cell consists of two electrodes (one zinc and the other copper) immersed in a solution of sulphuric acid, and connected by a conducting wire. The sulphuric acid ionizes into hydrogen (H^+) ions and sulphate (SO_4^{2-}) ions. Zinc ions (Zn^{2+}) enter the solution from the zinc electrode, leaving it with an excess of electrons. The excess electrons flow round the wire to the copper electrode, which thus acquires a negative charge. This attracts the hydrogen ions, which form hydrogen gas on the copper electrode.

67

Electricity and Magnetism

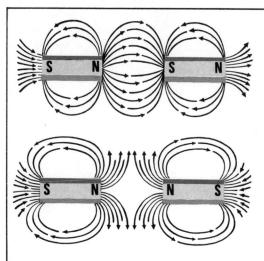

MAGNETISM

Magnetism gets its name from Magnesia in Asia Minor, where the ancient Greeks found a variety of iron ore (magnetite) which had the property of attracting pieces of metal to itself.

If a bar magnet is suspended from a thread it aligns itself north–south. This property is used in the compass. The south-seeking pole is its south pole, the north-seeking pole its north pole. If you hold two bar magnets with either both their north or both south poles close together, the magnets will move apart. If you place a south and a north pole close together the magnets will move towards each other. So like poles repel; unlike poles attract each other.

Around magnets are fields of force. These can be seen by sprinkling iron filings on a card resting on a bar magnet. The filings arrange themselves into a series of curved lines linking the north and south poles. These lines follow the so-called lines of magnetic force round the magnet, which make up its magnetic field.

d.c. Circuits

When a torch is switched on, an electric circuit —a path through which an electric current can flow—is completed, or made. The current flows around the circuit, which includes the fine metal filament of the bulb, causing it to heat up and glow brightly. As the current always flows in the same direction, it is called direct current (d.c.). This distinguishes it from alternating current (a.c.), which continually reverses its direction of flow.

All cells give rise to a direct current when their terminals are connected by a conductor of electricity. The current consists of electrons, which flow from the negative pole of the cell, through the conductor, and into the positive

Electricity and Magnetism

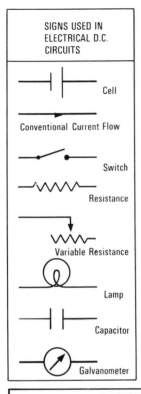

SIGNS USED IN ELECTRICAL D.C. CIRCUITS

Cell

Conventional Current Flow

Switch

Resistance

Variable Resistance

Lamp

Capacitor

Galvanometer

pole of the cell. A flow of 6·28 million million million electrons per second is equal to a current of one ampere.

Potential Difference

A potential difference (p.d.) is said to exist between two points if a current flows when they are connected by a conductor. Thus, a cell has a potential difference between its terminals. And a conductor carrying a current must have a potential difference between its ends. Like e.m.f., p.d. is measured in *volts* (V), and is often referred to as the *voltage* between the two points.

Ohm's Law

The strength of the current (I) that flows through a conductor with a potential difference (V) between its ends depends on two factors: the size of the potential difference and the extent to which the conductor opposes, or *resists*, the passage of current through it. This opposition to current flow is called the *resistance* (R) of the conductor. It is measured in units called *ohms* (Ω), after the German

RESISTORS IN SERIES AND PARALLEL

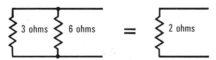

3 ohms   6 ohms   =   9 ohms

The total resistance of resistors connected in series is calculated by simply adding the individual resistances together. Therefore, if resistances of 6 ohms and 3 ohms are connected in series, the total resistance is 9 ohms.

3 ohms   6 ohms   =   2 ohms

The total resistance of resistors connected in parallel is calculated by using the formula $\frac{1}{R} = \frac{1}{R_1} + \frac{1}{R_2} + \frac{1}{R_3}$ ... etc.

In the example given $\frac{1}{R} = \frac{1}{3} + \frac{1}{6} = \frac{1}{2}$. Therefore R = 2 ohms.

Electricity and Magnetism

scientist Georg Ohm (1787–1854). By experiment, Ohm discovered that the ratio V/I is constant for some conductors under fixed physical conditions (such as temperature). This constant is the resistance of the conductor under those conditions. Hence, $V/I = R$. This relationship can be arranged to give $I = V/R$, or $V = IR$.

A *resistor* is a component which has the property of resistance. Its resistance may be fixed or variable.

Power

Power is the rate at which energy is converted from one form to another. Converting energy at the rate of one joule per second is equal to a power of one *watt*.

When a direct current flows through a conductor, electrical energy is converted to heat and, sometimes, light. The rate at which this happens—the power (P) in watts—is equal to VI, where V is the potential difference between the ends of the conductor, and I is the current passing through it.

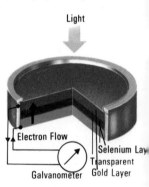

Below: A selenium photoelectric cell produces electricity from light. When light passes through the transparent gold layer and falls on the selenium, electrons are released. These flow around the circuit. The deflection of the meter needle depends on the intensity of the light falling on the cell. This arrangement is used in the photographer's light meter.

DEVELOPMENT OF THE ELECTROMAGNET

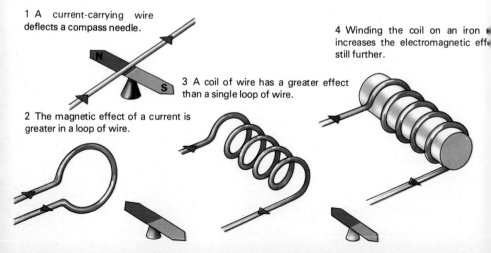

1 A current-carrying wire deflects a compass needle.

2 The magnetic effect of a current is greater in a loop of wire.

3 A coil of wire has a greater effect than a single loop of wire.

4 Winding the coil on an iron increases the electromagnetic effect still further.

Electromagnetism

Electricity is related to magnetism. The study of this relationship is termed electromagnetism. If you place a compass near a wire carrying an electric current, the compass needle will be deflected. The needle returns to a north-south position when the current is switched off. The passage of an electric current through the wire sets up a magnetic field around it.

The magnetic field around a wire carrying a current can be intensified by winding the

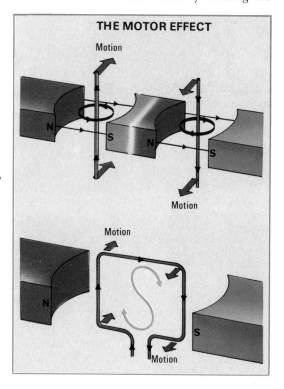

THE MOTOR EFFECT

If a current is passed through a wire in a magnetic field, it will tend to move. The direction of motion depends on the directions of the field due to the magnet and the field due to the current through the wire. On one side of the wire, the like field of the magnet repels the wire. On the other side, the unlike field of the magnet attracts the wire. The motor effect will cause a pivoted loop or coil of wire carrying a current to twist in a magnetic field. Moving coil meters and some electric motors make use of this effect. The simplest method of finding which way the loop will twist is to consider its magnetic poles. Here, the south pole of the loop must twist to face the north pole on the left.

Electromagnetism

wire into a coil. The coil then has the same kind of magnetic field as a bar magnet, but only when the current is passing. Such a coil is called a solenoid. The magnetism of a solenoid can be intensified by incorporating a piece of soft iron inside and around the coil. This forms the basis of the electromagnet. Small electromagnets are used in telephones and electric bells for attracting the armature that operates the bell striker. Powerful electromagnets are used at scrapyards for picking up iron and steel scrap.

Another electromagnetic principle can be demonstrated by passing a current through a wire placed near a magnet. The wire moves because of the interaction between its magnetic field and that of the magnet. This simple principle is exploited in the electric motor, which is widely used to drive machines in industry and to propel vehicles.

The reverse of this principle also holds. If you move a wire situated within a magnetic field, an electric current is set up, or induced in it. This phenomenon is electromagnetic induction.

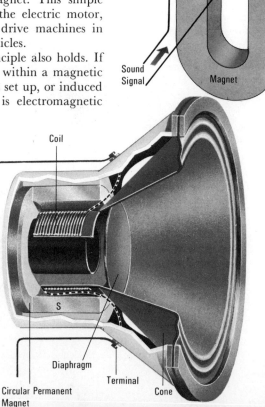

If a wire is moved between the poles of a magnet, a current will be induced in it. In the ribbon microphone, sound waves cause the metal ribbon to vibrate. This sets up a current in the ribbon and wire that is an electrical copy of the sound waves.

A moving coil loudspeaker uses the motor effect to convert electric signals into sound waves. The signals are fed to a coil, which is positioned between the poles of a circular permanent magnet. Variations in signal strength and direction cause the coil to move to and fro. This vibrates the attached cone, which emits sound waves corresponding to the varying currents in the coil.

Electromagnetism

THE DYNAMO EFFECT

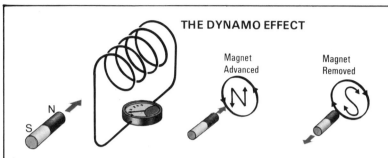

Moving a magnet towards or away from a coil causes a current to flow through the coil. By Lenz's law, the current induced in the coil opposes the motion causing it. When the north pole of a magnet is moved towards the coil, the current direction is such that it sets up a north pole to oppose the motion of the magnet. If the magnet is moved away, a south pole is set up and the current is reversed.

In a moving coil microphone sound waves make the diaphragm vibrate. This moves the coil to and fro in the field of the permanent magnet. An e.m.f. is thus induced in the coil. This e.m.f. is an electrical copy of the sound waves striking the diaphragm.

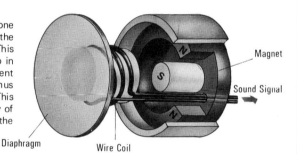

D.C. Electric Motors

In a simple d.c. motor, a pivoted coil of wire, wound on an iron core, is mounted in the field of a fixed permanent magnet. Connections from a battery to the coil are made via two fixed carbon blocks called *brushes*. These rub against a split metal cylinder called a *commutator*, each half of which is connected to one end of the coil. As in the moving-coil ammeter, the coil acts as an electromagnet. Passing a current through the coil magnetizes it, and it starts to turn to align its north and

Electromagnetism

south poles with the south and north poles of the permanent magnet. Here, the similarity with the ammeter ends, for the motor has no springs to restrict the coil's movement. So the coil turns freely to face the poles of the magnet. But, like any moving body, the coil cannot stop immediately, and its inertia carries it past the poles of the magnet. As the coil passes the alignment position, the *commutator* reverses its connections with the brushes, thus reversing the direction of the current through the coil. This reverses the magnetic poles of the coil, causing each one to be repelled from the fixed pole it has just passed and attracted to the other side of the magnet. As a result, the coil continues to turn in the same direction. The commutator reverses the current in this way every half-turn, so the coil rotates continuously.

The diagrams show the sequence of events during one rotation of a simple d.c. motor. For clarity, the coil has been shown as a single loop of wire. When a current flows through the coil, a magnetic field is set up, and the coil has a north and a south pole, represented here as a ghost magnet. The poles of the coil change places as the coil rotates, and so one end of the ghost magnet has been marked with a dot for easy identification. Separate permanent magnets provide a magnetic field in which the coil turns.

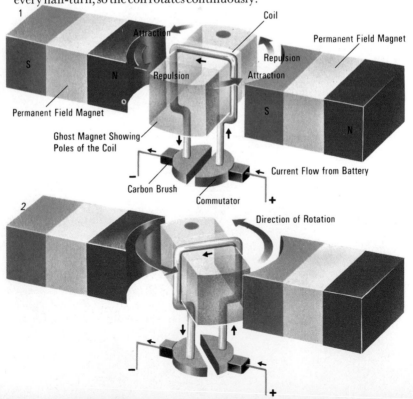

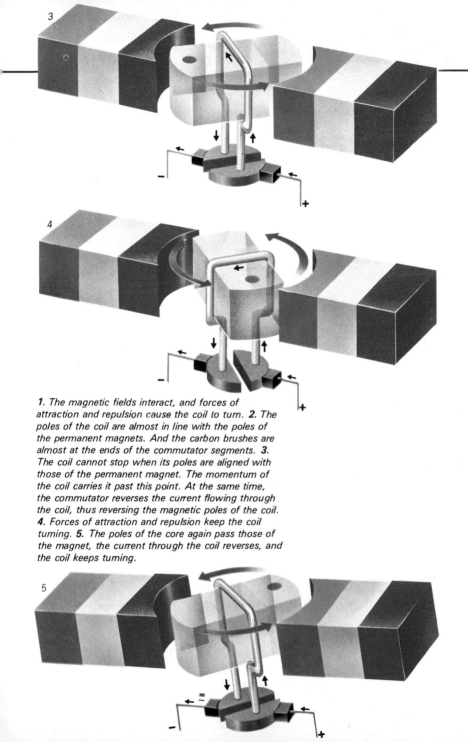

1. The magnetic fields interact, and forces of attraction and repulsion cause the coil to turn. *2.* The poles of the coil are almost in line with the poles of the permanent magnets. And the carbon brushes are almost at the ends of the commutator segments. *3.* The coil cannot stop when its poles are aligned with those of the permanent magnet. The momentum of the coil carries it past this point. At the same time, the commutator reverses the current flowing through the coil, thus reversing the magnetic poles of the coil. *4.* Forces of attraction and repulsion keep the coil turning. *5.* The poles of the core again pass those of the magnet, the current through the coil reverses, and the coil keeps turning.

Electromagnetism

Simple d.c. motors like this have limited use. One problem is that the torque, or turning force, produced varies considerably, being highest when the plane of the coil coincides with the direction of the field of the permanent magnet. To overcome this defect, the rotating part, called the *rotor*, or *armature*, can be wound with several coils. These are positioned evenly around the rotor and connected to a commutator with one pair of segments for each coil. This arrangement produces a high torque with little variation in strength.

The torque produced by an electric motor depends also on the strength of the field in which the rotor turns. By using a powerful electromagnet instead of a permanent magnet, the strength of the field and, hence, the torque developed by the motor can be greatly increased. Electromagnets used to produce a magnetic field are called the *field* coils or windings. As, in this case, they make up the stationary part of the motor, they are also referred to as the *stator*. In the *series motor*, the field and armature windings are connected in series with each other. In the *shunt motor*, the two windings are connected in parallel.

A moving coil ammeter is used to measure current, and a moving coil voltmeter is used to measure the potential difference across a resistance. The ammeter is a galvanometer that contains a low 'shunt' resistance. The voltmeter contains a high 'multiplier' resistance in series.

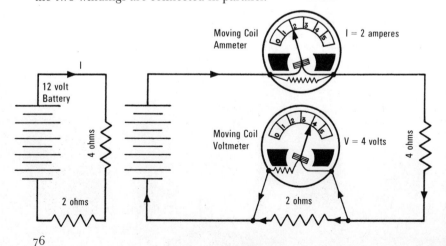

Electromagnetism

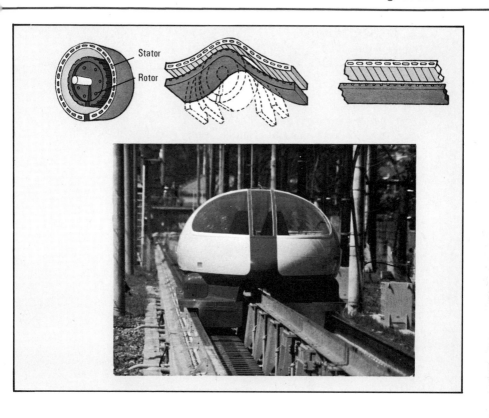

LINEAR INDUCTION MOTOR

A conventional motor produces rotary (turning) motion. It has a fixed outer *stator* and an inner *rotor*, which turns. A linear motor is designed to produce motion in a straight line. It may be thought of as an 'unwrapped conventional motor'. If such a motor is cut open and flattened out and the 'rotor' is then fixed, the 'stator' will be induced to move horizontally by the magnetic field.

This principle has been used in a number of designs, even as far back as the late 1800s. But the first linear motor to be built was a tubular d.c. motor that was used to launch missiles in 1917. Many people are now convinced that the linear electric motor could be used in railway systems. The photograph shows a prototype for a tracked hovercraft that uses a linear motor.

Linear motors have the advantage of having no parts that move against each other. Thus there is no need for lubrication and wear and tear is reduced.

Electromagnetism

A.C. Electric Motors

Motors designed for operation from an alternating current supply take many forms. But the most common type, by far, is the *induction motor*. Alternating current passed through fixed field coils sets up a rotating magnetic field. This moving field induces currents in closed loops of wire mounted on the rotor. These currents set up magnetic fields around the wires and cause them to follow the main magnetic field as it rotates. The operation of the induction motor depends on the rotating field passing through the loops on the rotor. Therefore the rotor must always turn more slowly than the rotating field. As no current has to be supplied to the rotor, the induction motor is simple to construct and reliable in operation.

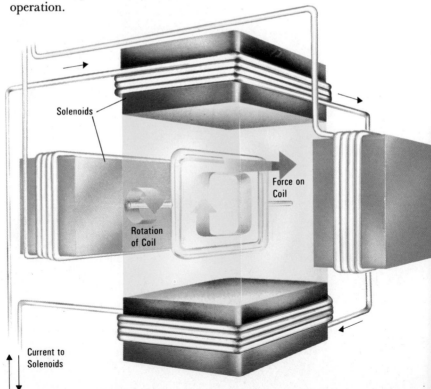

Electromagnetism

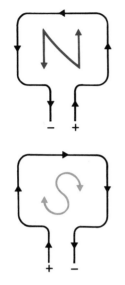

A quick way to find whether the end of a coil is a north (N) or south (S) pole.

How Generators Work

A simple generator resembles a simple electric motor. A coil, wound on an iron core and mounted so that it can rotate, is positioned between the poles of a permanent magnet. Rotating the coil makes it cut the magnetic field of the permanent magnet, so an e.m.f. is induced in the coil. The field can be thought of as imaginary magnetic field lines joining the poles of the magnet. As the coil rotates, the angle at which it cuts through these lines changes all the time. Therefore, the *rate* at which it cuts field lines changes too. This causes a corresponding variation in the strength of the induced e.m.f. The rotation of the coil also causes the induced e.m.f. to alternate, or change direction. This is because the coil windings alternately reverse the direction of their motion through the magnetic field. The result is an e.m.f. that increases from zero to a maximum, decreases to zero, increases to a maximum in the opposite direction, and then decreases again to zero.

In one type of generator, the ends of the coil are connected to metal cylinders called *slip rings*, which are mounted on the rotor. A

An induction motor is so called because the driving force is due to an electric current induced in a rotor, due to its interaction with a magnetic field. The motor in the diagram consists of four solenoids (coils wound on iron bars) and a closed loop of wire. When alternating currents are passed through the solenoids, they create a rotating magnetic field. This field cuts the closed loop, and thus induces an electric current in it. The loop therefore starts to turn. As it turns faster it tries to catch up the rotating magnetic field, and the difference between the two speeds gets smaller. The size of the induced current, and hence the size of the driving force, also get smaller. The loop of wire therefore settles down to a steady speed, which is lower than that of the rotating magnetic field.

Electromagnetism

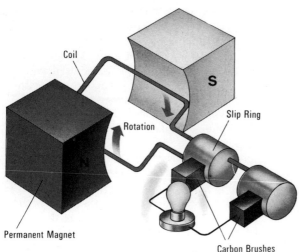

In a simple a.c. generator an alternating current is produced as the coil rotates in the magnetic field. By Lenz's law, the generated current sets up a magnetic field that tends to oppose the motion. Therefore, the face of the coil that is approaching the south pole of the magnet must also be a south pole. This decides the direction of the current flow. When the coil is horizontal, the maximum current is being generated, because the coil is cutting vertically across the magnetic field. As the coil reaches the vertical position, the current falls to a minimum. As it passes the vertical position the direction of current flow reverses.

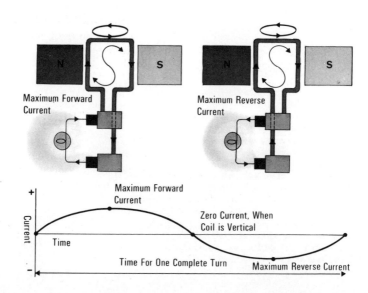

Electromagnetism

pair of fixed carbon brushes makes contact with these rings as the rotor turns. Therefore, the e.m.f. induced in the coil appears across the brushes. This alternating e.m.f. gives rise to an alternating current in a conductor connecting the brushes. For this reason, the generator described is called an *a.c. generator*, or *alternator*.

In a simple *d.c. generator*, the coil is connected to a commutator instead of slip rings. The current flowing through the coil still alternates. But the commutator reverses the connections to the carbon brushes every half-turn, so the current always leaves the coil through one particular brush and returns through the other one. Although the current supplied by such a generator always flows in the same direction through a conductor connecting the brushes, its strength varies as the rotor turns. This effect can be reduced considerably by arranging several coils around the rotor, each one being connected to a separate pair of segments on the commutator.

Today, only the smallest generators contain permanent magnets. In most generators,

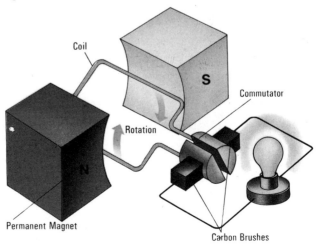

In a simple d.c. generator the current is generated in exactly the same way as in the a.c. generator. The current direction in the coil reverses each time the coil passes the vertical position. But in a d.c. generator, the connections of the split ring commutator also reverse, and the current direction in the circuit remains the same.

Electromagnetism

electromagnets are used because they can provide a much stronger magnetic field. In a.c. generators, these *field coils* are usually energized by a separate d.c. generator. Most d.c. generators, however, make use of an effect called *self-excitation* and need no separate supply for the field coils. A small amount of magnetism, known as *residual magnetism*, is always present in the iron cores of the field coils. This causes a small current to be generated when the rotor turns. Part of this generated current is passed through the field coils, increasing the strength of the field and, hence, increasing the generated current. In this way, the generated current builds up rapidly to full output.

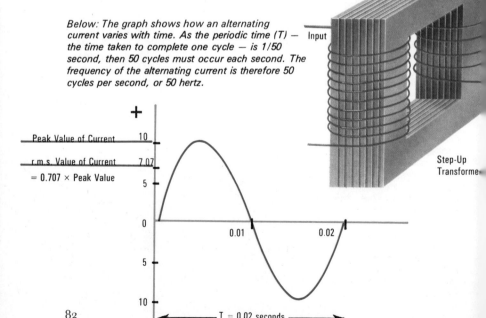

Right: Large transformers are used to convert the output of the generators in a power station to a high voltage and a low current. This is to reduce power loss in the distribution cables.

Below: The graph shows how an alternating current varies with time. As the periodic time (T) — the time taken to complete one cycle — is 1/50 second, then 50 cycles must occur each second. The frequency of the alternating current is therefore 50 cycles per second, or 50 hertz.

Step-Up Transformer

Alternating Current

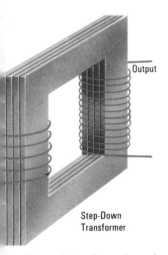

Above: A transformer is used to step-up or step-down an a.c. voltage. The alternating current in the input coil induces a changing magnetic field in the iron core. This induces an alternating current in the output coil. The final voltage depends on the number of turns in the output coil. Hence, a step-up transformer has more turns in the output coil than the input coil, and a step-down transformer has fewer turns in the output coil.

Alternating current (a.c.) changes all the time. It builds up to a maximum flow in one direction, decreases to zero, builds up to a maximum in the opposite direction, and then returns to zero once more. This complete sequence, or *cycle*, repeats, the rate at which it repeats being called the *frequency* of the alternating current. The mains current supplied to most houses in Europe has a frequency of 50 cycles per second. This is usually expressed as 50 hertz (50 Hz), the hertz being a unit equal to one cycle per second. The time taken for an alternating current to go through one complete cycle is called the *periodic time*, or *period*.

A current that continually changes its direction of flow might, at first, seem rather inconvenient. In fact, the opposite is true. Alternating current can be generated with greater efficiency than direct current. It can be easily transformed from one voltage to another, whereas direct current cannot. And alternating current can be easily converted to direct current, whereas the reverse process is much more complicated and less efficient.

Because an alternating current varies continuously, the problem arises as to how to describe its strength. One way is to state the maximum value it reaches. Another way is to state the *average value*. To have any useful meaning, the average value would have to be taken over one complete half-cycle, starting and ending at zero strength. Over a complete cycle, the average values would be zero.

Radio and Television

Radio was predicted before it was invented. In 1865, the Scottish scientist James Clerk Maxwell stated that it should be possible to make an alternating current send electromagnetic waves through the air. In 1887, an experiment by Heinrich Hertz proved Maxwell right. Hertz used a high alternating voltage to produce sparks. Each time a spark occurred, a pulse of current passed through the circuit, and electromagnetic waves were radiated. The waves produced a spark across the gap in a nearby brass ring. This was the first deliberate transmission and detection of radio signals. Like light, X-rays, and all other forms of electromagnetic radiation, radio waves travel at 300 million metres per second.

Hertz could detect the radio waves only when his receiver—the brass ring—was close to the transmitting apparatus. But other scientists, fascinated by the prospects of sending messages from one place to another without using connecting wires, soon developed much more sensitive detection systems. The most sensitive of the early detectors was the *coherer*. It consisted of a glass tube containing metal filings between a pair of electrodes. Radio waves made the filings cohere, or stick together, allowing a current to pass between the electrodes. In one type of

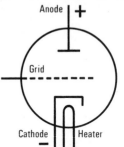

Above: The triode valve has a wire mesh, called a grid, between the cathode and the anode. The charge applied to the grid affects the electron flow from cathode to anode. For example, a high negative charge on the grid causes it to repel any approaching electrons (also negative), so that none reach the anode. A small, varying voltage applied between grid and cathode causes larger corresponding variations in the electron flow through the valve and associated components. In this way, the triode amplifies the signal applied to the grid.

Right: Heinrich Hertz (1857-1894) used an induction coil (a type of transformer) to produce a high voltage. This caused a spark to jump between the knobs of the transmitter, and radio waves were emitted. The waves were picked up by the receiving ring and caused a small spark to jump across the gap.

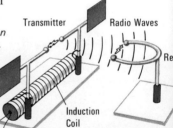

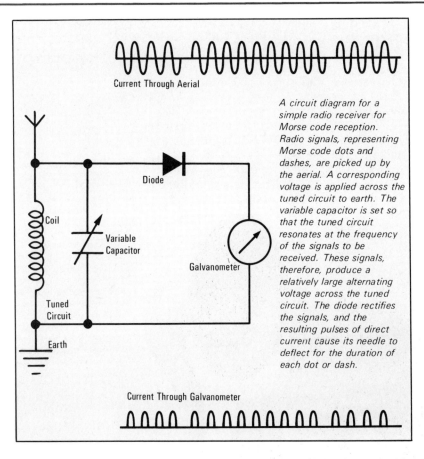

A circuit diagram for a simple radio receiver for Morse code reception. Radio signals, representing Morse code dots and dashes, are picked up by the aerial. A corresponding voltage is applied across the tuned circuit to earth. The variable capacitor is set so that the tuned circuit resonates at the frequency of the signals to be received. These signals, therefore, produce a relatively large alternating voltage across the tuned circuit. The diode rectifies the signals, and the resulting pulses of direct current cause its needle to deflect for the duration of each dot or dash.

radio receiver, the coherer was connected in series with a battery and electric bell. When radio waves were received, the filings would clump together, completing the circuit and making the bell ring.

The Italian Guglielmo Marconi developed practical systems for radio communications in the 1890s. Marconi achieved great success and, in 1901, his radio station in Cornwall transmitted a signal that Marconi received in Newfoundland.

Radio and Television

By this time, tuned circuits, consisting of a coil and capacitor, were being used to tune in, or select, radio signals picked up by a wire aerial. But one great problem remained. The coherer, although sensitive, was not reliable in operation. Sir John Ambrose Fleming solved the problem in 1904, when he introduced the diode (two-electrode) valve for detecting radio signals. This valve was a practical application of the *Edison effect*, discovered by Thomas Edison in 1883. While experimenting with electric lamps, Edison had discovered that a small current would

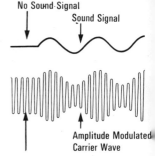

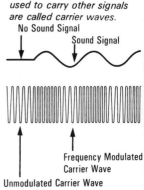

In amplitude modulation (above), the amplitude, or strength of the waves, is made to change. In frequency modulation (below), the sound signals vary the frequency of the radio waves. Radio waves used to carry other signals are called carrier waves.

An early radio with triode valves, headphones and loudspeaker.

Radio and Television

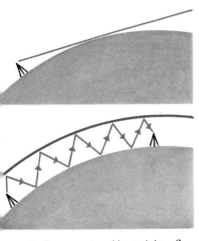
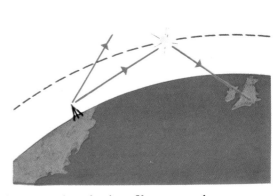

Radio waves travel in straight lines. Ground waves (top) cannot be used over long distances because of the curvature of the earth. Other waves travel upwards until they hit the ionosphere, a layer of ionized gas, from which they are bounced back to earth. Some radio waves bounce up and down between the ionosphere and earth until they circle the planet. These waves can be received anywhere on earth. But not all radio waves are reflected from the ionosphere. Some very short waves such as those used for television go straight through and disappear into space. To send television signals across the Atlantic, they have to be bounced back to earth from a special satellite.

flow between the glowing filament and a separate metal electrode placed in the evacuated bulb. The important feature of the diode was that it passed current in one direction only. An alternating current produced in an aerial by a received radio signal could not be detected on a galvanometer. For the needle could not follow the rapidly alternating current, and the average value of the current was zero. But the diode could be used to *rectify* the signals—convert them to pulses of direct current. And the pulses could deflect a galvanometer because, as the current flowed in one direction only, its average value was no longer zero.

In 1906, it was found that some crystals act as rectifiers when a fine wire is placed in them. So, like the diode valve, the crystal diode could be used to detect radio signals. But a much more important discovery was made in the same year. In the United States, Lee de Forest found how to make a valve amplify. He simply placed a fine wire grid between the electrodes of a diode, thus making it into a *triode*, or three-electrode valve. Small voltage variations applied to the grid cause large, corresponding variations in the current passing through the valve. De Forest's invention was vital to the further development of radio and other branches of electronics.

Early radio communication was carried out by transmitting bursts of radio waves in Morse code. Then, it was realized that, if the strength, or amplitude, of a radio signal could be *modulated* (made to vary) according to the strength of a sound signal produced by a microphone, it would be possible to transmit speech. By the end of World War I (1914–18), radiotelephony (voice communication by radio) was common. In a typical transmitter, a valve circuit called an *oscillator* generated radio signals. Other valves amplified the speech signals and produced a corresponding varying voltage. This was used to alter the strength of the radio signals produced by the

Tuned Circuits

In radio receivers, a mixture of signals picked up by an aerial passes through a tuned circuit. This is used to select, or tune in, signals of one particular frequency. Each signal passing through the circuit produces voltages across the coil and capacitor.

As the impedance of these components varies with frequency, the voltages across them will depend on frequency too. Around one particular frequency called the *resonant frequency* of the circuit, the voltages are relatively large. This is the frequency selected by the circuit. The voltages produced by signals of other frequencies are usually so small that they do not interfere with the selected signal. Normally, the capacitor is variable. Altering its value changes the resonant frequency and, therefore, the station selected. Although all coils have some resistance, this is not normally shown on circuit diagrams.

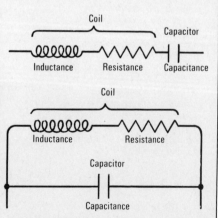

A tuned circuit contains a coil and a capacitor, either in series (top) or in parallel (bottom).

Radio and Television

A variable capacitor (sometimes called a condenser) is used to select or 'tune into' a radio station that is broadcasting on a particular frequency. When the plates of the capacitor are in the correct relative positions, the tuned circuit resonates to the frequency of the selected radio station.

Right: A transistor is formed from crystals of semiconductor material, such as germanium or silicon. A slice of crystal is treated with certain impurities to form three regions with distinct electrical properties. These regions are called the emitter, base, and collector. A small, changing current flowing between the base and the emitter causes large variations in the current flowing between the emitter and collector. In this way the transistor can amplify signals.

oscillator. In the receiver, a diode or triode detector demodulated the received signal, separating from it the required sound component. This was then reproduced on headphones or, after amplification, on a loudspeaker.

Regular radio broadcasting to inform and entertain the general public started in the 1920s. At first, most people listened on headphones connected to simple crystal sets, for receivers of this kind were simple to operate and required no power supply. But gradually, as the number of radio stations increased, much more sensitive and selective valve receivers, with several tuned circuits and amplifying stages, became popular. In the 1930s, a new system for sound transmission was introduced. Instead of the sound signals making the strength of the radio waves (amplitude modulation) it modulated their frequency instead (frequency modulation). As the receiver had to detect frequency changes, not amplitude changes, it did not reproduce any amplitude changes caused by electrical interference. As a result, almost all background noise was eliminated, giving excellent reception.

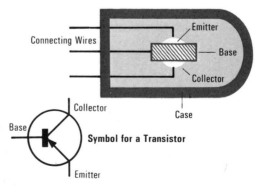

Television

The discovery of the photoelectric effect in the 1870s led many scientists to attempt to transmit pictures by wire. A lens could be used to form an image of a scene, and an array of photoelectric cells could be used to convert the brightness of each part of the image into corresponding voltages. These voltages would then be used to reproduce a picture of the scene, perhaps by illuminating an array of tiny electric lamps. Unfortunately, the power produced by the cells was too small to illuminate a lamp, so the system was impracticable. The development of the amplifying triode valve in the early 1900s made such a system possible. But it was inconvenient to use hundreds of pairs of wires to carry the vision signals from the array of photoelectric cells. To avoid this, a technique called *scanning* was used. The image to be transmitted was scanned, or sampled, bit by bit. Signals corresponding to the brightness of each element were transmitted, one after another, along a single pair of wires. In England in the 1920s, John Logie Baird used a rotating scanning disc in a system designed to transmit silhouettes. The disc had been invented by the German scientist Paul Nipkow in 1884. An image of the scene was formed on the rotating disc. Holes in the disc passed, one at a time, across the image, exposing a photoelectric cell

Below left: In the image-orthicon camera tube, the camera lens forms an image of the scene on a light sensitive cathode. This emits a pattern of electrons. These electrons strike the target and dislodge electrons from it, leaving it with a charge pattern, or 'electrical image'. The electron beam from the cathode sweeps quickly across the target, line after line. At each point on the target, the beam deposits just enough electrons to replace those previously dislodged. The returning beam therefore varies in strength, according to the charge pattern on the target. This varying beam is used to form the television signal.

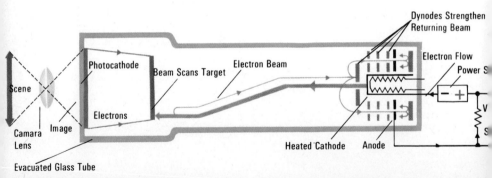

Radio and Television

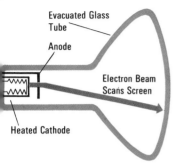

The television picture tube is a special form of cathode-ray tube that converts the vision signal into an image of the televised scene. The vision signal controls the strength of the electron beam as it scans the fluorescent screen, line by line. The screen glows according to the strength of the moving beam. Hundreds of lines of varying brightness together form a completed picture.

behind the disc to each part of the image in return. Baird amplified the signals produced by the cell and used them to vary the brightness of a small neon lamp. A second scanning disc rotated in front of the lamp. As each hole in the disc passed quickly across the flickering lamp, it appeared to the eye to form a line of varying brightness. A complete image was built up in this way from a series of lines, scanned quickly, one after another. A mechanical scanning system was used for the first regular experimental television broadcasts, which were made in New York in 1928. The vision signals were made to modulate radio waves, in the same way as speech signals were transmitted by radio. Several other television stations started to broadcast using mechanical scanning systems. But these were gradually replaced by the more reliable electronic scanning system, which had been developed in the United States by the Russian physicist Vladimir Zworykin. In this system, modified cathode-ray tubes were used in the camera and receiver. In the camera tube, an electron

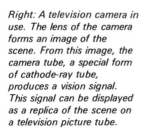

Right: A television camera in use. The lens of the camera forms an image of the scene. From this image, the camera tube, a special form of cathode-ray tube, produces a vision signal. This signal can be displayed as a replica of the scene on a television picture tube.

Radio and Television

beam was made to scan an image of the scene and produce a vision signal. In the receiver's picture tube, the image was reconstructed by an electron beam scanning a fluorescent screen. This is the system used today.

The most common type of black-and-white camera tube used in broadcasting is the image-orthicon. An image of the scene is formed on a layer of photoelectric material, causing it to emit electrons in a pattern corresponding to the light values in the scene. These electrons strike a glass disc called a target plate, knocking electrons from it and

Below: John Logie Baird, one of the great pioneers of television, standing beside his transmitting machine in the world's first television 'studio'.

Radio and Television

Below: Viewing a transmission in a BBC studio in the 1930s when television was a novelty, almost a toy.

leaving it with a pattern of positive charges. Rapidly changing magnetic fields make the electron beam in the tube scan the 'electrical image' on the target, line by line. The beam deposits just enough electrons (negative charges) to cancel the positive charge at each point on the target. So the electron beam leaving the target varies in strength according to the number of electrons 'lost' to the target. The variations in beam strength are amplified to form the vision signal.

In a television picture tube, the vision signal is made to vary the strength of an electron beam as it scans a screen of fluorescent material. The screen glows according to the strength of the beam at each point. Therefore, as the beam strength varies, the brightness of the line it traces out varies too. The brightness pattern thus formed on the screen corresponds to the brightness pattern of the image formed on the camera tube. So the picture tube reproduces the original scene.

Although the electron beam in the picture tube strikes only one point on the screen at a time, it moves so rapidly that, to the eye, it appears as if the whole screen is glowing continuously.

In order to reproduce the scene correctly, the beam in the picture tube must scan exactly in time with the beam in the camera tube. For this reason, timing signals called synchronizing pulses are added to the transmitted vision signal. In the receiver, these pulses control the scanning of the electron beam.

Colour Television

In one type of colour television camera, optical devices form three separate images containing the red, green, and blue colour components of the scene. The three coloured images are scanned by camera tubes to form

Radio and Television

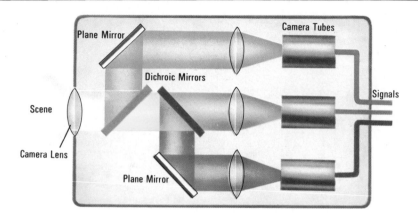

Above: A colour television camera contains tinted mirrors, called dichroic mirrors, which split up the light from the scene into blue, green, and red colour components. The images are then scanned by three image-orthicon camera tubes, and three separate vision signals are produced. These are then combined in a special way, so that they can be sent along a single cable, and transmitted like a black and white signal.

Below: An early television set, with the picture seen in a mirror in the lid.

three vision signals. These signals are then combined in a special way that enables them to be separated again, when this is required. The combined signal is transmitted in the normal way.

In the receiver, the colour components are separated and fed to a picture with three electron beams. These scan the tube together, forming red, green, and blue images respectively. Each colour signal controls the strength of the corresponding beam, so three superimposed colour images are formed. We see these images as a single, full-colour replica of the scene being televised.

Baird's televisor, shown here in its 30-line 1930 version, lost the battle with its rival, the Marconi-EMI electronic television invented by Zworykin. But Baird was the first man to show true television pictures in 1926.

Below: In a colour television receiver, the three colour components are separated out from the combined vision signal. Each colour component controls the strength of an electron beam. The three beams scan the screen, which is coated with minute dots of chemicals called phosphors in a particular arrangement (far right). These glow blue, green, or red when struck by electrons. The perforated mask ensures that each beam strikes only one type of phosphor. As a result, three superimposed colour images are formed on the screen. We see these images as a single, full colour picture of the scene being televised.

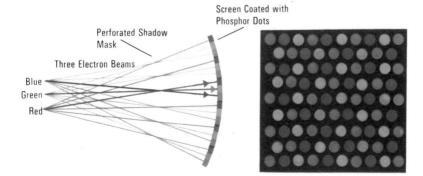

Computers

Digital Computers

Computers are tools that enable us to calculate quickly and conveniently. Digital computers perform calculations involving *counting*. When ancient man began to acquire possessions, he needed a counting system. To start with he used his fingers. But large numbers were difficult to manage in this way, and counting systems that used groups of units and tens were devised. These led to the development of the abacus. In the early abacus, pebbles were placed in columns between lines marked on a table or other smooth surface. Even today, the abacus is still the most common calculating machine in the world. In the modern abacus, beads are slid along wires to represent counts of units, tens, hundreds, thousands, and so on.

In 1642, a French mathematician called Blaise Pascal invented a new kind of calculating machine. Pascal's machine used a system of toothed wheels, each with ten teeth. Numbers could be added by turning the wheels by an appropriate number of teeth.

By the 1800s, many other mechanical calculators had been invented. Some worked well. Others, although theoretically sound, failed because their delicate mechanisms could not be manufactured with sufficient accuracy. The most outstanding designer in this field was Charles Babbage, a mathematics professor at Cambridge University, England. In 1832, Babbage invented the first automatic general purpose computer. It was called an *analytical engine*, and it had many features that are used in modern electronic computers.

BINARY CODE

The binary system of numbers uses zeros and ones only. The system uses a 'yes/no' principle—a one indicates 'yes' and a zero indicates 'no'. The binary system can therefore be used to show the presence or absence of certain decimal numbers. In practice, these numbers are 1, 2, 4, 8, 16, 32, 64, and so on, as shown below.

```
64 32 16 8 4 2 1
 0  0  0 0 0 0 1 =  1
 0  0  0 0 0 1 0 =  2
 0  0  0 0 1 0 0 =  4
 0  0  0 1 0 0 0 =  8
 0  0  1 0 0 0 0 = 16
 0  1  0 0 0 0 0 = 32
 1  0  0 0 0 0 0 = 64
```

The binary number 10010101 therefore equals the decimal number 149 (128 + 16 + 4 + 1).

Binary numbers can be punched onto tape, which can then be fed into a computer. On punched tape a hole indicates a one, or 'yes', and no hole indicates a zero, or 'no'. On the piece of punched tape shown below the following numbers have been punched:
1011 = 11
 100 = 4
1100 = 12
 111 = 7
1001 = 9
1101 = 13

More than one century elapsed before the first successful automatic general purpose computer was built. This machine was completed in 1944 by Howard Aiken of Harvard University, Massachusetts. Aiken's calculator contained counter wheels, electric motors to turn them, electromagnets, and many other parts—over 750,000 in all.

Electronic Digital Computers

ENIAC (Electronic Numerical Integrator and Calculator) was designed by electrical engineer J. Presper Eckert and physicist John Mauchly. Completed in 1946, ENIAC was the first automatic electronic digital computer. More than 150,000 watts of electricity were needed to power the circuits, which contained about 18,000 valves, 1500 relays (switches operated by electromagnets), and hundreds of thousands of other components. Numbers were fed into the machine on punched cards. The coded holes in the cards were used to form pulses of electricity, which were then fed to electronic circuits that carried out the calculations.

In this abacus, the numbers 48 and 75 are being added together. First, 1748 is set on the abacus (A). The number 5 is then added to the units column (B). But this would give 13 in the units column, so 1 is added to the tens column, leaving 3 in its column. Finally, the number is added to the tens column. (C) Again, this would give 12 in this column, so 1 is added to the hundreds column, leaving 2 in the tens column. The final result is therefore 1823.

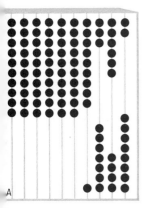
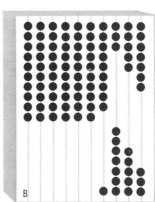
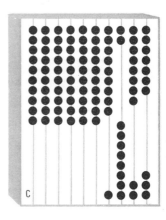

Computers

A cash register is a mechanical digital computer. It contains wheels that turn to certain positions to represent numbers. Addition is accomplished by counting the number of turns that a wheel makes. Modern cash registers are usually electrically operated.

For most purposes, valves were considered to be extremely reliable. But ENIAC had so many of them that failures were frequent. In later computers, the valve was replaced by much more reliable, efficient, and smaller solid-state devices—first the transistor, and then the integrated circuit.

Modern electronic digital computers handle numbers in *binary code*. This is because the binary system uses only two digits: 0 and 1. These digits can be conveniently represented in the computer by the absence (0) or presence (1) of electrical pulses, holes in a card, or magnetic spots on a tape or disc.

Before it can be used to solve a problem, the computer must be given instructions about what to do. The instructions, which form the *program*, are held in coded form by magnetic rings called *cores*. These form the computer's *store*, or memory. The *data* (figures to be used in the calculation) are fed into the computer from an *input device*. For example, the data may be fed straight into the computer from a keyboard machine operated like a typewriter. Or a tape containing the information in code may be played into the computer. Electronic circuits in the *processor*, or arithmetic unit, carry out the calculation, the whole sequence of operations being directed by the *control unit*. The results of the calculations are obtained on *output devices*, which include cathode-ray-tube screens and high-speed printers.

Analog Computers

Analog computers calculate by *measuring* quantities similar, or *analogous*, to the quantities being calculated. The mercury thermometer, for example, is a simple analog computer for calculating temperature. As the temperature varies up or down, the height of the mercury column varies in a similar manner. The height of the column, as measured on a scale marked in degrees, gives the temperature.

In electronic analog computers, electrical quantities represent the information being dealt with. For example, a simple circuit could be used to work out the speed of a vehicle, given the distance travelled and the time taken. The formula for this calculation is:

$$\text{Speed} = \text{Distance} \div \text{Time}$$

This is easy to represent electrically because

$$\text{Current} = \text{Voltage} \div \text{Resistance}$$

If the voltage across a resistor is used to represent distance, and the resistance is used to represent time, then the current through the resistor will represent speed.

Below: A printed circuit board being inserted into a computer. The computer is built up using many such boards, which can easily be removed. This aids the location and repair of faults.

Computers

Computers

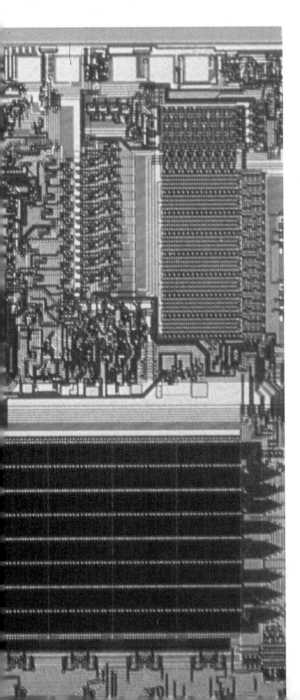

Some silicon chips carry out only one aspect of computing, such as arithmetic operations. Others have the highly complex circuitry of a complete computer and can process and control information; these are 'computers on a chip' or microprocessors. The microprocessor shown here, enlarged 50 times, has different areas devoted to memory (the two large rectangles), and central processing (the rectangle at bottom left) which can perform more than 300,000 logical functions per second. The most advanced chips are capable of performing millions of logical operations per second.

Sound

IF A HUGE boulder crashes down a mountainside in some remote part of the country and there is no one there to hear it, is any sound produced? One answer to this teaser is 'no' – because sound is something we hear and if it is not heard then there is no sound! But the answer can also be 'yes', if we define sound simply as vibrations made by an object – by the boulder crashing.

All sounds from a tiny whisper to the roar of a jet plane have one thing in common. They are vibrations which, when they travel through the air and reach our ears, we hear as sound. You can feel sound vibrations. For example, gently touch a bicycle bell as it is ringing.

Sound vibrations travel in waves in all directions from the source of the sound. If you throw a stone in a pool, you will see ripples spreading out from the point where the stone hit the water. Sound waves travel in much the same way. A vibrating object makes the surrounding air vibrate.

Sound waves are transmitted by air molecules vibrating back and forth, but the molecules do not travel with that wave. This is similar to what happens in a water wave. The ripples spread out in circles from the source, but the water particles – and anything in the water – simply bob up and down as the wave passes. There is a basic distinction between a water wave and a sound wave. In a water wave the water particles vibrate up and down at right angles to the direction in which the wave is travelling. It is called a transverse wave. In a sound wave the air particles vibrate back and forth along the direction of travel of the wave. It is called a longitudinal wave. The strength

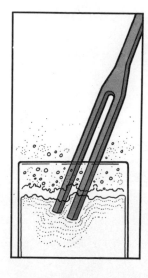

All sounds are vibrations. If you dip a tuning-fork in a tumbler of water the vibrating fork will splash water out of the tumbler.

Sound

of sound waves – their intensity – is measured in decibels.

Anything that vibrates produces sound. A slammed door or a rusty gate hinge makes an unpleasant, irregular sound called noise. Noise can be irritating, distracting, and even harmful. By contrast the vibrations of a piano's strings can produce very pleasant sounds. They are regular, musical sounds.

In human beings the voice box, or *larynx*, produces sounds. When we speak, air from the lungs passes through the larynx and causes two small bands of tissue called the *vocal cords* to vibrate.

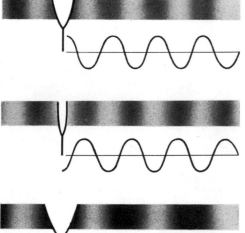

Sound waves, alternate compressions and rarefactions of air, are formed when a tuning fork vibrates.
As the prongs of the tuning fork move outwards, a compression is formed.

As the prongs move inwards, a rarefaction forms.

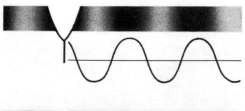

If the fork is struck harder, the compressions are greater, and the sound is louder.

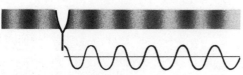

If a smaller fork is struck, the vibration (frequency) is more rapid and the wavelength is shorter; a higher-pitched sound is heard.

Sound

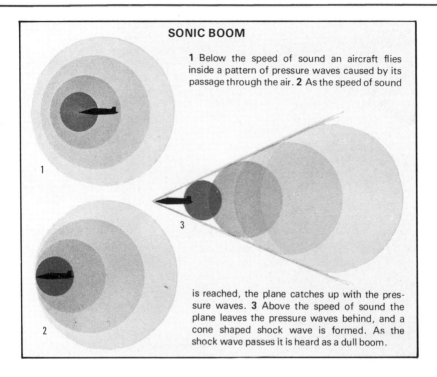

SONIC BOOM

1 Below the speed of sound an aircraft flies inside a pattern of pressure waves caused by its passage through the air. **2** As the speed of sound is reached, the plane catches up with the pressure waves. **3** Above the speed of sound the plane leaves the pressure waves behind, and a cone shaped shock wave is formed. As the shock wave passes it is heard as a dull boom.

Acoustics

Acoustics is the branch of science and engineering concerned with sound, particularly with sound transmission and its control. An acoustics engineer in a concert hall tries to ensure a clear and faithful transmission of musical sound to the audience. In a factory or office he is usually concerned with preventing the transmission of unwanted noise from voices and machinery – often very difficult to do.

An acoustics engineer designing a concert hall has to control the way sound is reflected and absorbed. If the sound is reflected too much, disturbing echoes may be set up. If it is absorbed too much, the sounds will die away too quickly giving music a 'dead' quality.

Sound

Sound recording

Sounds can be recorded on discs, or on magnetic tape, or on film for reproduction in the cinema. The first component in any sound-recording system is the microphone. This device converts the pattern of sounds to be recorded into corresponding electrical impulses. These impulses must then themselves be converted into some form in which they can be preserved. To play back the recording, the 'preserved' electrical impulses must be regenerated and fed into the last component in the sound-recording system, the loudspeaker. This recreates the original sound pattern.

The first recording of sound was carried out by converting the microphone signals into mechanical vibrations. This is the method still used for disc recording. The microphone impulses are made to vibrate a stylus or needle that cuts a pattern as it spirals round the surface of a coated disc. To play back the sounds on a record player, a stylus is inserted into the spiral groove formed in the disc. As it follows the impressed pattern, the stylus vibrates. The vibrations are converted back into electrical impulses and fed to a loudspeaker.

In these days of mass production single discs are seldom cut. The first stage of commercial disc recording is, oddly enough, to make a good tape recording of the music, dialogue, or whatever is to be recorded. The tape is then played back through special apparatus which vibrates a cutting stylus on the lacquered surface of an aluminium 'master' disc. Duplicates of it are taken, and from them numerous 'stampers' are made. The commercial discs are produced from the stampers.

Both monoaural (mono) and stereophonic (stereo) records are made. In stereo recording microphones are so positioned that they pick up 'left-hand' and 'right-hand' sounds. The

Below: The first phonograph, the invention of Thomas Alva Edison. The recordings were made on cylinders.

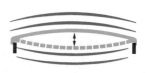

Drums sound when a stick or mallet strikes the taut skin and sets it vibrating (above). The note may be changed by altering the tension of the skin.

Sound

STEREOPHONIC SOUND

A stereophonic record has two signals recorded separately on each side of the groove. When the needle of a stereophonic pickup passes along the groove, each side applies a different component of motion (x and y) to the stylus. At the top of the pickup are two piezoelectric crystals. Crystal A responds to the y component of motion, and crystal B responds to the x component. The crystals produce electric currents which then pass to two separate amplifiers that are linked by the same controls; volume, treble, bass, and balance. The amplified currents then pass to two loudspeakers. Each loudspeaker reproduces the sound originally recorded on one side of the groove. The listener, placed in front of and between the speakers, hears the combined sound.

Sound

slightly different signals produced are recorded on opposite sides of the record groove. On playback the signals from the different sides of the groove are fed to separate loudspeakers. The speakers reproduce 'left-hand' and 'right-hand' sounds respectively, which gives a greater sense of depth and realism.

Thomas Edison invented the forerunner of the record player, the phonograph, in 1877. At first sounds were recorded on waxed cylinders. Discs came into use in the 1880s.

The tape recorder

The tape recorder preserves sounds on magnetic tape. The pattern of electrical signals coming from the recording microphone is converted in the tape recorder into a pattern of magnetic signals. This magnetic pattern can then be preserved on a tape coated with iron oxide crystals, which act as miniature magnets. Chromium oxide tapes are also widely used, especially in cassette recorders. Valdemar Poulsen pioneered development of the tape recorder in the early 1900s.

In a typical tape recorder the tape is drawn from one spool to another past record and playback 'heads'. The heads are small electromagnets which become magnetized when electric currents pass through them. In recording, the electrical signals from the microphone are passed through the coils of the record head. They set up corresponding magnetic signals that align the iron-oxide particles on the tape in a certain way.

On playback, the magnetized tape passes the playback head and sets up in its coils corresponding electrical signals. These are exact replicas of the microphone signals that were originally fed to the record head. When they are fed to the loudspeaker, the original sounds are reproduced.

Right: In tape recording sound is converted into electrical impulses by the microphone. The recording heads, small electromagnets which become magnetized when electric current is passed through them, convert the impulses into magnetized patterns on the iron-oxide coating of the tape. To replay, the tape is passed across the playback head and the magnetic patterns are converted back into electric impulses.

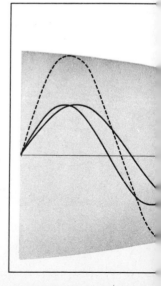

Sound

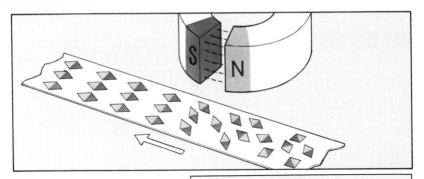

RESONANCE

An instrument can be made to sound without playing it. If a note reaches it that is in tune with any part of the instrument free to vibrate, then the instrument itself will sound as well. Open the piano and press down the loud pedal so that the strings are free to vibrate. Then sing loudly into the piano, while still pressing the pedal. When you stop singing, you will hear the piano ringing with sound; the strings have been set vibrating by the sound of your voice. Many instruments make use of this effect, which is called resonance. The body of a violin or guitar resonates with the sound of the strings and amplifies the sound.

BEATS

When any two notes sound together, one also hears the difference of their frequencies. This does not become apparent until the notes are close in frequency—that is, one is slightly out of tune with the other. Then their difference will be heard as a slow rise and fall in volume. This effect is known as beats. Listening for beats is a very good way of tuning an instrument.

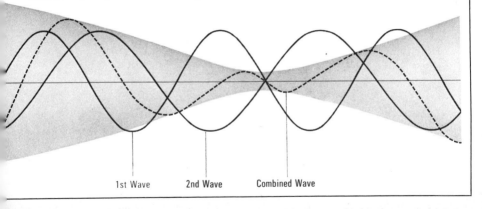

1st Wave   2nd Wave   Combined Wave

Heat and Temperature

Heat, like so many things in life, is something that we cannot do without but also something of which we can have too much. A human being or any animal is a delicate balance of energy conversions. Too much heat—or too little—upsets the balance and puts life in danger. The amount of heat that reaches the Earth from the Sun is enough to support life over most of the Earth's surface. But no life exists without protection at the Poles; it is so cold that there is not enough heat energy to maintain the energy balance.

Heat and Temperature
It is important to realize that the temperature of anything does not measure the amount of heat that it possesses. Temperature is a measure of the degree of motion of the molecules. A hot object has faster-moving molecules than a cold object. The amount of heat is related to the mass of an object as well as its temperature. A large mass of a substance at a low temperature could have a greater amount of heat than a small mass of the same substance at a high temperature.

The motion of molecules gives rise to heat — and to cold. Heat is one of the most important forms of energy, and is put to great use by man. Here an engineer cools some piston linings before inserting them in an engine block. On insertion, the linings warm and expand, fitting them tightly in place.

Heat and Temperature

Several temperature scales have been devised. The Centigrade scale places the freezing point of water at 0°C and the boiling point at 100°C. It is also known as the Celsius scale after the Swedish physicist Anders Celsius who first proposed the scale in 1742 (though he suggested that freezing point be 100° and boiling point 0°, an odd decision that was soon reversed). Temperatures below freezing, which often occur in winter, are shown as negative or minus figures in the Celsius scale. To avoid this, the Dutch physicist Gabriel Fahrenheit had earlier proposed the Fahrenheit scale, in which the freezing point of water is 32°F and boiling point is 212°F. In 1848, the British physicist Lord Kelvin pointed out that the logical starting point of any temperature scale is absolute zero. This temperature should be 0° and all other temperatures should be measured from it in degrees equal in magnitude to degrees Celsius. This scale, called the Kelvin scale or Absolute scale, is now

Clinical Thermometer

Above: The clinical thermometer, like ordinary thermometers, works by expansion. It has a bulb containing mercury that expands along a narrow bore in a calibrated glass tube as the temperature increases. The clinical thermometer contains a constriction which prevents the level of the mercury falling immediately the thermometer has been removed from the patient's mouth to be read. Ordinary thermometers often contain coloured alcohol instead of mercury. These glass thermometers have a limited range.

THE RANGE OF TEMPERATURE

The range of temperatures that are to be found in the Universe is enormous — from as much as 20 million kelvins (K) at the centres of stars to near absolute zero (0 K) in outer space. But all other temperatures are very near the lower end of such a scale. The Sun's surface has a temperature of 6000 K and iron melts at 1812 K. Paper catches fire at 557 K and water boils at 373 K. The highest recorded temperature on Earth is 332 K and the lowest 185 K, while water freezes at 273 K. Air liquefies at about 75 K.

Heat and Temperature

used in science and the SI unit of temperature is called the kelvin (K). Thus absolute zero is 0K, the freezing point of water is 273·16K and the boiling point of water is 373·16K.

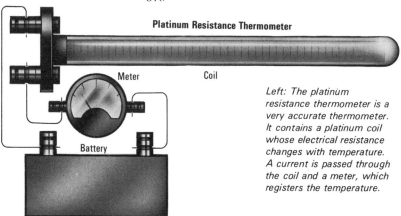

Platinum Resistance Thermometer

Left: The platinum resistance thermometer is a very accurate thermometer. It contains a platinum coil whose electrical resistance changes with temperature. A current is passed through the coil and a meter, which registers the temperature.

Below: The optical pyrometer measures the colour of light coming from a red-hot or white-hot source to find its temperature. This is done by comparing the colour of a filament heated by an electric current to the colour of the source. When the filament seems neither brighter nor darker than the source (right), the meter is read to find the temperature of the source.

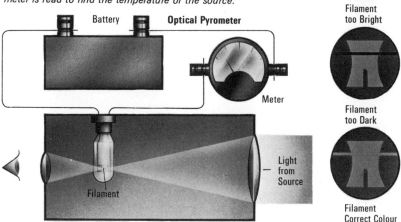

Filament too Bright

Filament too Dark

Filament Correct Colour

Heat and Temperature

Heat is energy and it is measured in joules (J). The amount of heat in an object depends on its mass and its temperature. It takes twice as much heat to raise the same mass of a substance by twice the temperature (in kelvins), or to raise twice the mass of substance by the same temperature. But the final temperature depends on its specific heat capacity. A piece of copper will gain 50K in temperature when it receives the same amount of heat that produces a rise of only 5K in the same mass of water. Copper therefore has a specific heat capacity ten times that of water. Specific heat capacity is measured by giving a known mass of a substance a known amount of heat and measuring its rise in temperature.

THE WORLD'S WINDS

Set patterns of winds occur throughout the world. The trade winds blow towards the equator, but at the equator it is calm. At temperate latitudes, the prevailing winds come from the west, and winds also blow away from the poles. These general movements of the atmosphere are convection currents produced by the Sun's heat reaching the Earth. This is greatest at the equator, and causes air to rise as it is warmed. Colder air moves in to replace it, producing the trade winds. The rising air spreads out on each side of the equator, becoming colder and eventually descending again and returning to the equator. Similar convection currents produce the westerlies and the polar winds. The Earth rotates beneath them as they move, causing them to veer in direction.

Convection also produces on-shore gusts at the seaside by day and off-shore breezes by night. During the day the land absorbs the Sun's heat more readily than the sea. Air therefore rises above the land, and cooler air blows in from the sea to replace it. At night the land cools more quickly than the sea. Air therefore rises above the sea and winds flow from the land towards the sea. (See below).

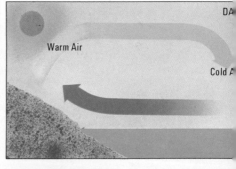

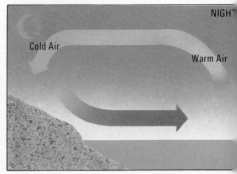

Heat and Temperature

Above: The fennec fox lives in hot desert regions, where the temperature is so high that it cannot quickly rid itself of body heat. Its huge ears, however, allow the heat to be dispersed over a wide area. They act like cooling fins on an engine.

In SI units, it is expressed as the number of joules of energy required to give a mass of 1 kilogram of the substance a rise in temperature of 1K.

Conduction

If you have ever burnt your fingers by picking up a hot pan from the cooker, you have experienced conduction. Conduction is the flow of heat through solids. If one end of a solid object is heated, the molecules at that end vibrate faster. They set their neighbouring molecules vibrating more, and these molecules jostle their neighbours and so on. Heat gradually moves through the object. Heat spreads from the hot plate or gas flame throughout the pan and, as you pick it up, heat flows into your fingertips and produces a

Heat and Temperature

sensation of burning. Similarly, if you pick up a piece of ice from the refrigerator, heat flows from your fingertips by conduction into the ice, and you sense cold.

Most metals are good conductors of heat, particularly copper and aluminium. Saucepans and kettles are mostly made of aluminium, which is also cheap and light. Wood, plastics and cloth are poor conductors. The handles of pans and kitchen utensils are made of plastics or wood because these materials will hardly even become warm at the temperatures used for cooking. We encase ourselves in clothes to stop body heat from leaving our bodies (though some gets through, otherwise we would boil), and we insulate water heaters and roofs in houses to cut down heat losses and save fuel costs. Extra layers of fat beneath the skin similarly provide insulation in polar animals such as whales and seals.

Liquids also are poor conductors (apart from mercury, which is a metal). So too are gases. The fur and feathers of animals trap layers of air next to the skin so that heat does

Above: The ancient Greek engineer Hero, who lived during the first century AD, invented a famous steam engine that bears his name. In Hero's engine, steam from a boiling kettle is fed through pipes to a hollow sphere with two vents on opposite sides. The sphere is mounted so that it can rotate. Steam rushes from each vent, pushing the sphere in the opposite direction and making it spin.

CONVERTING TEMPERATURES

To change Fahrenheit to Celsius (Centigrade), subtract 32 and multiply by 5/9. For example, 68°F = 20°C (68 − 32 = 36; 36 × 5/9 = 20). To change Celsius to Fahrenheit, multiply by 9/5 and add 32. For example, 10°C = 50°F (10 × 9/5 = 18; 18 + 32 = 50). To change Celsius to Kelvin, add 273.16; to change Kelvin to Celsius, subtract 273.16. In practice, whole numbers are often used and the 0.16 of a degree is dropped.

Left: A thermostat contains a bimetallic strip of iron (above) and brass (below). When cool, the strip is straight; the contact is closed (left) and current flows through it to a heater. As the temperature rises, the brass expands more than the iron, making the strip bend. At one stage contact is broken and the current stops flowing (right), cutting off the heater. When the bimetallic strip cools down, contact is made again and the current flows once more.

To Heater

To Heater

Heat and Temperature

not escape. This is why ducks can swim happily in freezing water. We ourselves are similarly insulated because our clothes, particularly woollen garments, trap air. Double glazing keeps the heat in because the layer of air between the two panes of glass prevents conduction through the window. However, heat transfer in liquids and gases is complicated by movement of the liquid or gas.

Convection

If you place your hand over a radiator, you will feel a lot of heat rising from it, even though not much heat is to be felt in front of it. The hot, rising air eventually spreads throughout the room, bringing warmth to every corner. This kind of heat transfer

Below: Rivets are applied hot, so that they contract on cooling and produce a strong, tight join.

Heat and Temperature

Above: As a beaker of water is heated, it takes up heat and its temperature rises (1 and 2). When it reaches boiling point, it begins to boil (3). More heat is needed to keep it boiling, but the temperature does not increase (4). A lot of heat is needed just to change water to steam. This heat is called latent heat.

Right: Ice melts if it is compressed, because melting points vary with pressure. Two blocks of ice can be made to stick together simply by pressing them. Water forms where they meet, because the pressure lowers the melting point of ice, and then refreezes as the pressure is released. Ice skaters momentarily cause the ice to melt beneath their skates, enabling them to skim over the ice.

is known as convection. The radiator—or any other heater placed in a room—warms the air next to it. The air expands as it gets hotter and becomes less dense than the surrounding air. It therefore rises from the heater and colder air comes in to replace it. This too is heated and rises, and a current of warm air is circulated throughout the room. This air movement is known as a convection current. Convection also occurs in liquids. As a kettle is heated, convection currents in the water automatically stir it so that heating occurs evenly throughout the water. When it boils, all the water is at boiling point.

Radiation

If you stand in front of an open fire, you will feel the warmth soaking into you. Conduction cannot be transferring the heat, because there is air between you and the fire and air is a poor conductor of heat. But convection cannot occur either, because the air heated by the fire is escaping up the chimney. The heat is reaching you by radiation; heat rays produced by the fire move through the air and warm everything that absorbs them. All

Heat and Temperature

objects produce heat rays, the amount depending on their temperature.

The heat involved in changes of state is called *latent heat* (meaning hidden heat), because it seems to come from or go to nowhere. Evaporation is an unusual change of state in that it can take place over a wide range of temperature. Other changes of state —freezing, melting and boiling—take place at fixed temperatures. As a change of state occurs that requires heat—melting or boiling —latent heat is taken up by the solid or liquid to make the change. The temperature of the substance being heated remains constant while the change takes place. Latent heat must be removed from a liquid to produce freezing; the temperature drops until freezing begins and then remains steady as it takes

Heat and Temperature

place. This explains why it takes such a long time to make ice in a refrigerator. While the water cools quickly, it freezes only slowly because it gives out latent heat as it does so. The refrigerator takes away this extra heat as fast as it can.

The production or absorption of latent heat produces a fixed melting or freezing point and a fixed boiling point for every substance. For solid-liquid change, the energy change is called the latent heat of fusion and for liquid-vapour change, it is called the latent heat of vaporization. The amount of latent heat involved in a change of state is surprisingly large. It takes nearly as much heat to melt a piece of ice as it does to raise the same mass of water from freezing point to boiling point. And it then requires over five times as much heat to change the boiling water to steam.

Condensation—the change from vapour to liquid—is like evaporation in reverse. It occurs when the air is saturated with water vapour and can hold no more.

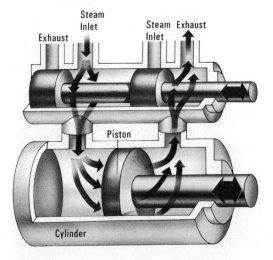

In the reciprocating steam engine, a piston is forced up and down a cylinder by steam pressure. Ports open and close alternately to admit high-pressure steam to one side of the piston while opening an exhaust on the other side. Steam locomotives are powered by reciprocating engines.

Putting Heat to Work

There are two basic kinds of heat engines: external combustion engines and internal combustion engines. In external combustion engines, heat is produced outside the engine and taken to the chamber where work is produced. For example, steam is raised in a boiler and then conducted to a steam engine. In internal combustion engines, heat is produced inside the working chamber, generally called the combustion chamber. Petrol and jet engines are internal combustion engines, as is the rocket.

The power of all heat engines depends on the temperature at which heat is put to use in the engine, and the temperature at which it leaves the engine. In between, some heat will have been converted into work and so the greater the temperature difference (in kelvins), the greater the work done. In a petrol engine, this temperature difference is

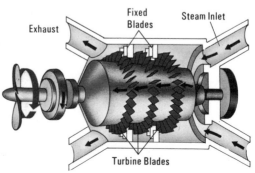

In a steam turbine, high-pressure steam is directed through groups of fixed blades to strike the blades of a series of turbine wheels. The steam expands as it passes through each set of turbine blades, driving the wheels round. The fixed blades direct the steam onto the turbine blades at the correct angle. Ships and electric generators are powered by steam turbines.

Putting Heat to Work

that between the temperature of the burning fuel and the temperature at the exhaust and at the surface of the engine. No engine uses up all its heat. In the petrol engine, only about a third is used; two-thirds has to be rejected into the atmosphere.

The Steam Engine

The first steam engines were slow, ponderous machines. They worked on the principle that as the steam in a chamber condenses, it produces a partial vacuum. In the first practical steam engine, invented by the British engineer Thomas Savery (1650–1715) in 1698, the vacuum was used to suck up water from mines; it had no moving parts other than valves. Thomas Newcomen (1663–1729), another British engineer, utilized the same principle to drive a piston up and down a cylinder and produce mechanical energy. Like all heat engines, these early steam engines worked by a cycle of operations repeated over and over again to obtain continuous power. However, the cycle was impossibly slow, because the chamber containing the steam had to be cooled every time to produce condensation of the steam inside and then

Right: A heat exchanger is used to transfer heat from a source of heat to a place where it is needed. The source of heat is a hot liquid or gas, such as steam, and it circulates in a chamber through which runs a pipe containing another liquid or gas that needs to be heated. Here, a heat exchanger is used to transfer heat from hot steam produced by a nuclear reactor to a plant for distilling fresh water from sea water. The cold sea water is heated in the heat exchanger, producing water vapour that condenses on the cold inlet pipe to give fresh water.

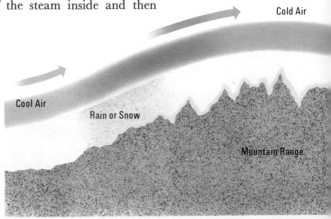

Putting Heat to Work

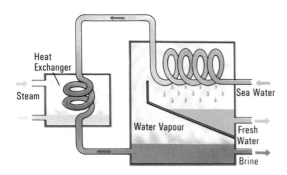

heated to admit the steam without it condensing. James Watt (1736–1819), another British engineer, shortened the cycle time in 1769 by connecting a condenser to the steam chamber. Steam was admitted to both the chamber and condenser, but the condenser was kept cold and the chamber hot. As the steam condensed in the condenser, it produced a partial vacuum in the condenser and the steam chamber; there was no need to cool and heat the engine alternately and it would drive the piston up and down as fast as its power allowed.

Modern steam engines do not use pistons and cylinders, but contain rotating sets of turbine blades. The high-pressure steam spins

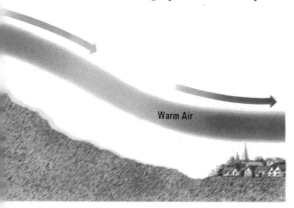

The snow covering plains lying in the lee of mountains often melts unexpectedly as warm winds descend from the snowy heights of the mountains. These winds are known as föhn winds in Europe and chinook winds in America. As the wind meets a mountain range, it ascends and the air becomes cooler. It sheds any moisture it has as rain or snow on the mountain slopes and gains some latent heat. But as the air rises further, it expands adiabatically and cools. The dry, cold air then blows over the snowy summits of the range and descends rapidly to the plains on the other side. As it does so, it is compressed and becomes much warmer. Föhn winds aid agriculture by melting winter snows. But they can also cause avalanches to occur in the mountains.

Putting Heat to Work

the blades of the turbine just as a wind rotates a windmill. As the steam passes through the blades, it expands and loses pressure and heat. The steam turbine was developed by the British engineer Sir Charles Parsons (1854–1931) in 1884. It was much more useful than a piston engine, in which the power is produced as a reciprocating (up-and-down) motion, because the rotary power can go directly to the propellers of a ship or the shafts of electric generators in power stations, the two principal uses of the steam turbine.

The Petrol Engine

In the petrol engine and other internal combustion engines, ignition of the fuel inside the combustion chamber causes the air and gases produced by burning of the fuel to expand violently. The force of this expansion is then turned into useful power. In the petrol engine, it usually drives a piston up and down a cylinder and a crankshaft driven by a connecting rod beneath the piston converts this motion into rotary motion. Drive shafts take the rotary motion to the wheels through the clutch and gears. Most petrol engines contain more than one cylinder to give more power.

A petrol engine is a complex machine. The pistons move up and down the cylinders several thousand times a minute and with each movement fuel has to be admitted, ignited at a precise instant, and the exhaust gases removed before more fuel is introduced. The simplest petrol engine works by a two-stroke cycle; that is, all these operations take place within one upstroke of the piston and one downstroke.

The two-stroke engine is comparatively simple and cheap, but it does not develop great power and its rapid action uses up fuel quickly. It is used in motorcycles, but motor-

THE FOUR-STROKE CYCLE

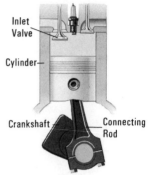

On the first stroke, or induction stroke, fuel is admitted through the inlet valve at the top of the cylinder as the piston moves down.

On the second stroke, or compression stroke, the inlet valve closes, and the piston moves upwards, compressing the fuel vapour.

Putting Heat to Work

Sparking Plug

On the third stroke, or power stroke, the sparking plug fires and the fuel ignites, forcing the piston down the cylinder.

Exhaust Valve

On the fourth stroke, or exhaust stroke, the exhaust valve opens and the piston moves up the cylinder, expelling the exhaust gases.

cars contain a four-stroke engine for power and economy. In this engine, fuel and exhaust gases enter and leave the combustion chamber through valves. On the first downstroke of the piston or induction stroke, fuel enters the cylinder. On the first upstroke or compression stroke, it is compressed. Ignition then occurs and expansion of the gases in the cylinder occurs on the second downstroke, the power stroke, and on the second upstroke, the exhaust stroke, the exhaust gases leave. The cycle then begins again. As power is produced only during the power stroke, a flywheel is connected to the crankshaft to keep the pistons moving during the other strokes.

The Jet Engine

However, a rotary internal combustion engine of very high power also exists in the gas turbine. This engine was developed by the British engineer Sir Frank Whittle (1907–) in World War II. Its high power makes it most suitable for aircraft, in which it is better

A Ford assembly line of 1914. Henry Ford was the first to make use of the principles of mass production.

125

Putting Heat to Work

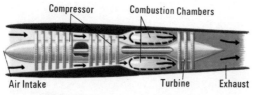

In a turbojet engine, a compressor draws air into the engine and compresses it. Fuel is sprayed into the combustion chambers and burns in the air. The gases produced expand rapidly, driving a turbine that is connected to the compressor. The jet of gases that leaves the engine acts to thrust it forwards.

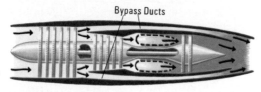

In a bypass turbofan engine, the compressor has fan-like blades to draw in more air, some of which is diverted around the combustion chambers to the exhaust. This action produces less noise and more thrust, making the engine useful for airliners.

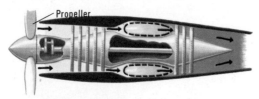

In a turboprop engine the turbine also drives a propeller as well as the compressor. Some thrust is also developed to power the aircraft.

The ramjet is the simplest form of jet engine. It has no compressor, and the air is 'rammed' into the engine by the forward motion of the engine. The ramjet can therefore come into action only when it is already moving.

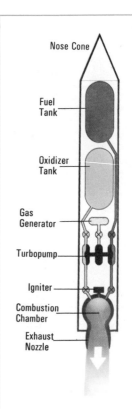

Above: A liquid-fuel rocket contains tanks of fuel and oxidizer that are pumped to the combustion chamber and ignited.

known as the jet engine. Air is drawn into the engine by a compressor, is compressed, and enters combustion chambers where it mixes with the fuel. Expansion occurs as the fuel burns and the hot expanding gases power a turbine before leaving the engine by the exhaust pipe. The turbine is connected to the compressor and drives it. In most engines, motive power is produced as a reaction to the motion of the exhaust gases; the expansion of these gases in the engine thrusts it forward as they move backward and out of the exhaust. The rotary power produced serves only to drive the compressor blades. However, in turboprop engines, the rotary power also goes to propellers to drive the aircraft, and comparatively little thrust is developed.

The ramjet is an even simpler form of engine, consisting of an open tube with a fuel supply. When it is moving, air rushes into the tube, mixes with burning fuel, expands and leaves the exhaust, producing a thrust.

The Rocket Engine

The rocket is as simple as the ramjet. It consists of a combustion chamber where two fuels mix and ignite. The expanding gases produced immediately rush from the exhaust, propelling the rocket in the opposite direction. No air is needed; the rocket can therefore work in space and its main use has been in spaceflight. Any fuels can be used that burn with each other. The most powerful rockets use liquid fuels—kerosene or liquid hydrogen together with liquid oxygen—that are pumped from separate tanks into the combustion chamber. Liquid-fuel rockets can be shut down and refired at any time. Solid-fuel rockets are simpler in design and are used where less control is necessary, as in automatic missiles.

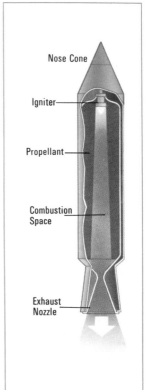

Above: A solid-fuel rocket contains fuel and oxidizer in solid form. The propellant has a hole through its centre along which the gases produced by combustion flow to the exhaust nozzle.

Colder than Cold

The second law of thermodynamics states that heat will, of its own accord, flow only from a hot object to a cold object. Despite appearances to the contrary, the action of a refrigerator conforms to this law. The basic action is simple. A liquid called a refrigerant moves through a closed pipe. When it reaches the interior of the refrigerator, it evaporates inside the pipe. Evaporation takes up latent heat, and the liquid cools as it evaporates. Heat flows from the refrigerator compartment into the vapour in the pipe. This vapour then moves to a condenser on the outside of the refrigerator and condenses to a liquid. Latent heat is given out and lost to the surrounding air from the surface of the condenser. The liquid then moves back into the refrigerator to evaporate again (see page 119). The interior of the refrigerator is kept cold overall because heat is taken from there to the outside. It is possible to work this method in reverse and take heat from outside to heat a house. This heating system is much more economical than using electrical heating. However, at no point does heat actually flow from a colder to a hotter place, as this is impossible. It flows into the cold vapour and away from the warm condenser.

Much lower temperatures can be reached by other methods. If a gas is alternately compressed and expanded, it will gain and lose heat. If it is allowed to lose the heat gained by compression but not to gain heat when it expands, it will lower its temperature adiabatically as it expands. Expansion also produces cooling because the molecules in the

Above: A man shatters a rubber tyre after it has been dipped in liquid air. At such a low temperature — about −200°C — many materials that are normally flexible become brittle.

gas lose energy in overcoming the attraction between them. By repeating these operations in a closed cycle, the gas will eventually liquefy. Liquid air is made in this way.

The science of very low temperatures is known as *cryogenics*, and strange things happen in this colder than cold world. Helium remains liquid even near absolute zero unless it is compressed, but it loses all viscosity and becomes a *superfluid*. It will flow up the walls of a container and over the top. Metals and alloys lose all electrical resistance and become *superconductors*.

Chemicals

VAST QUANTITIES OF chemicals are used in the modern world. Soaps and polishes, dyes, acids, artificial fibres, and explosives – all these things and thousands more are products of the chemical industry.

Acids such as sulphuric acid and alkalis such as caustic soda (sodium hydroxide) are examples of basic chemicals. They are the starting point in the manufacture of more complex chemical substances. Some products made from basic chemicals but requiring further processing are termed intermediate chemicals. Synthetic resins are in this group; they need processing to turn them into plastics, which are an example of finished chemicals. Most of the familiar products of the chemical industry are finished chemicals.

The modern chemical industry had its origins in the late 1800s. One of the most notable discoveries of that period was of coal-tar dyes by William H. Perkin (1856). This launched the modern dyestuffs industry. Six years later Alfred Nobel invented dynamite, launching the explosives industry. The plastics industry had its roots in the discoveries of Celluloid (nitrocellulose) by Hyatt in 1869 and of Bakelite by Baekeland in 1909.

Coal tar was the only important source of organic chemicals until the 1920s. Then petroleum began to be used as a commercial source, and today most countries rely on it.

Petroleum and coal tar contain mainly chemicals called hydrocarbons, made up of hydrogen and carbon only. They are called organic chemicals because they come from things that once lived. Organic chemicals contrast with chemicals such as sulphuric acid and sodium chloride (salt).

Right: A familiar chemical laboratory scene, which has not changed greatly over the years. But increasingly in research laboratories such sophisticated apparatus as mass spectrometers (below) are being used for analysis.

Chemicals

Chemical processing

The chemical industry takes many kinds of raw materials – minerals, air, wood, coal, petroleum, water – and turns them into useful products. Any chemical process – no matter what raw materials are used or products formed – can be broken down into a number of basic stages, or operations.

Many of the physical operations involved in chemical processing, such as mixing and dissolving, are simple in principle, but they are nevertheless very important in bringing a mixture or solution into a uniform state. Crushing and grinding involves the use of mills to reduce the size of materials. Filtering removes solid matter from a liquid suspension.

The sulphur burner in a sulphuric-acid manufacturing plant. Sulphuric acid is the most important industrial chemical and is often called the 'lifeblood' of industry. Over 80 million tonnes of the acid are produced in the world every year.

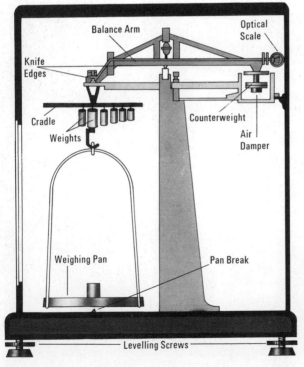

A modern single-pan balance, now widely used in chemical laboratories. It is a constant-weight type. With nothing in the pan, the weight of the weights and the pan just balances the counterweight. The sample to be weighed is placed on the pan, and weights are lifted mechanically from the cradle until the beam is in balance again. The weight of the sample is thus measured by the sum of the weights removed from the cradle. This contrasts with the traditional two-pan balance in which weights are added to one pan to balance the weight of the sample in the other.

Chemicals

Chemicals

Evaporation and drying removes moisture by means of heat or by vacuum. Distilling purifies or separates liquids by heating them until they vaporize and then cooling them so that they condense, or become liquid again.

Among the most common chemical operations is oxidation, which at its simplest means adding oxygen. Burning sulphur in air is an oxidation process that results in a gas – sulphur dioxide. Reduction is an operation in which oxygen (or its equivalent) is removed from a substance. Iron ore (iron oxide) is reduced to iron by reaction with coke (carbon) in a blast furnace.

Polymerization is another important chemical operation, by which many small molecules (monomers) are joined together to form a very long molecule (polymer). The common name for polymers is plastics. The gas ethylene, for example, can be made to polymerize and form the plastic polyethylene, or Polythene. Cracking – the reverse of polymerization – is the breaking down of large molecules into smaller ones, as in the breaking down of heavy oils into light petrol.

Electrolysis, or the splitting of a chemical compound by means of electricity, is a widely used tool in chemical processing.

Suitable conditions for chemical reactions may mean the application of heat, or pressure, or both. A catalyst – a substance that promotes a chemical reaction but is not itself changed by that reaction – may also be needed.

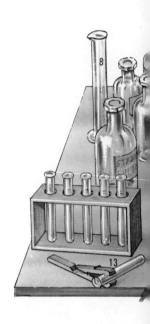

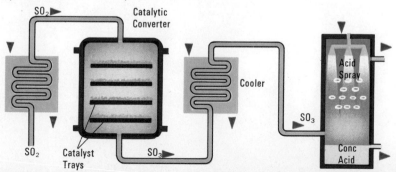

Chemicals

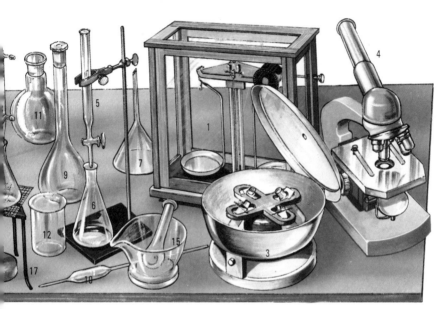

The illustration above shows some of the pieces of apparatus used in a chemistry laboratory. 1. Balance. 2. Reagent bottles. 3. Centrifuge. 4. Microscope. 5. Burette. 6. Conical flask. 7. Funnel. 8. Measuring cylinder. 9. Volumetric flask. 10. Pipette. 11. Flat-bottom flask. 12. Beaker. 13. Test-tube holder. 14. Bunsen burner. 15. Mortar and pestle. 16. Coil condenser. 17. Tripod stand.

The diagram on the far left outlines the contact process for sulphuric-acid manufacture. A mixture of sulphur dioxide and air is passed over a heated catalyst, whereupon the sulphur dioxide is converted to sulphur trioxide, which is led into a dilute acid spray. It reacts with the water present to form more acid, and concentrated acid collects at the foot of the tower. The catalyst for the contact process is usually platinum, finely divided in asbestos (platinized asbestos), or vanadium pentoxide.

The sulphur dioxide may be obtained in a variety of ways, depending on the cheapest material available to the acid-making plant. Occasionally elemental sulphur is burned in air. Often iron pyrites — iron sulphide — is roasted in air, whereupon sulphur dioxide forms.

Chemicals

Heavy chemicals
Basic chemicals such as sulphuric acid, caustic soda, and ammonia are used in very great quantities in the modern world and are called heavy chemicals. They contrast with other products of the chemical industry – for example, vitamins – that are produced in relatively small quantities and which are termed fine chemicals. Other important heavy chemicals include the mineral acids nitric, hydrochloric and phosphoric, and soda (sodium carbonate).

Sulphuric acid is probably the most important chemical produced by the industry. It is used on a vast scale in chemical processing and for many other purposes. It is used in car batteries and for 'pickling' steel to clean it.

Caustic soda, or sodium hydroxide (NaOH), is as strong an alkali as sulphuric acid is an acid. And like sulphuric acid, it can cause serious burns – hence its name 'caustic'. It is a

---

Branches of Chemistry

Like all sciences, chemistry grows as more ideas are tested and accepted. Scientists find that they become better experts if they specialize – that is, concentrate on one type of problem. The following are the chief branches of chemistry:

Analytical chemistry Experiments to find out which elements, and how much of each, make up any substance.
Applied chemistry Using chemistry to solve the problems of industry, medicine, agriculture, and other fields.
Biochemistry Studies living processes and helps fight disease.
Inorganic chemistry Study of substances that do not contain carbon.
Microanalysis Experiments with very small quantities of substances.
Organic chemistry Study of substances that contain carbon, such as sugars, alcohols, and many fuels.
Physical chemistry Studies chemical processes using properties of physics (such as energy, electricity, and radiation) and mathematics.
Polymer chemistry Study of long, chain-like molecules that contain many atoms.
Qualitative analysis Experiments to find the kind of chemicals (elements and compounds) in a substance.
Quantitative analysis Experiments to find the amounts of each chemical in a substance.
Structural chemistry Study of how atoms are arranged and the bonds between them.
Volumetric analysis Experiments with measured volumes of liquid substances.

Chemicals

very powerful cleaning agent – it is used in oven cleaners, for example. But one of its main uses is in the manufacture of soap. In soap making it is boiled with an oil or fat.

Ammonia (NH_3) is another essential industrial chemical. It is used to make fertilizers such as ammonium sulphate and ammonium nitrate. It is made into nitric acid, which is used on a large scale by industry in general and to make explosives such as TNT (trinitrotoluene) and nitroglycerine.

Most caustic soda these days is produced by the electrolysis of brine (sodium chloride). Chlorine is produced at the same time. The picture below shows the chlorine cell room.

Chemicals

Plastics

The plastics sector of the chemical industry has probably expanded more than any other in the past few decades. A great many plastics are in common use, each having different properties from the others which make it more suitable for different applications. Although the plastics have very different properties they do have two things in common. They can be shaped readily when heated, and they all possess long-chain molecules – molecules consisting of a chain of repeated small units. There may be many thousands of units in the chain. Such substances are called polymers, from the Greek word *poly*, meaning many.

The common plant substance cellulose is a naturally occurring polymer. It is not in itself a plastic because it cannot be readily shaped. Treatment with nitric acid yields a more suitable plastic material – cellulose nitrate. But cellulose nitrate is far too brittle to be useful until the 'plasticizer' camphor is added to it to produce Celluloid.

Synthetic plastics are made entirely from chemicals. Most of these common plastics are made from petroleum products.

Melting and setting

Plastics may react to heat in one of two ways. Some simply soften and melt, and return to a solid state when they cool. Others remain quite rigid. The heat-softening type of plastic is called a thermoplastic, the heat-resistant type a thermosetting plastic. In a thermoplastic the long chain molecules are free to move in relation to one another. But in a thermosetting plastic the long chains are cross-linked to form a rigid network.

The most common way of making plastic products is by moulding. In injection moulding thermoplastic chips are first heated until

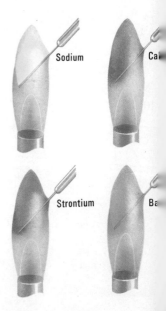

Below: The characteristic colours displayed by compounds of the alkali and alkaline-earth metals when thrust into a Bunsen flame during the flame test.

Sodium Ca

Strontium Ba

Chemicals

Filaments of viscose rayon, made from the natural polymer cellulose, leaving the acid bath. In the process of making viscose rayon, wood pulp is first steeped in caustic soda, pressed damp dry, and ground up to a form resembling breadcrumbs. This is then reacted with carbon bisulphide to form cellulose xanthate, which is dissolved in diluted caustic soda to form viscose. This is forced through tiny holes in a 'jet' immersed in an acid bath to form filaments of viscose rayon yarn. From the acid bath, the filaments pass up round a rotating wheel, and then down to join all the fibres from the other jets in a 'tow' (on the right hand side of the picture).

they melt, and then forced into a water-cooled mould under pressure. This method is ideally suited to mass-production techniques and is very widely used. In blow moulding air is blown into a blob of molten plastic inside a hollow mould, forcing it against the walls of the mould.

Chemicals

Thermoplastics can also be shaped by extrusion and calendering. In extrusion, molten plastic is forced through a shaped hole, or die. Fibres for textiles and sheet plastic may be made by extrusion, using different shaped dies. Calendering is a kind of mangling process which presses and squeezes softened plastic into a thin sheet.

Thermosetting plastic products can be made by compression moulding. In this method a moulding powder – a thermosetting resin which has not yet cross-linked into a rigid plastic – is placed in the lower half of a mould. The mould is heated and the resin starts to cross-link. Before this reaction is complete the top half of the mould is forced down on top of the lower one. The resin is squeezed into shape between the two halves. Laminating produces the heat-proof laminate used, for example, for the working surfaces in kitchens. Sheets of filler material such as paper and cloth are soaked in resin solution and placed one on top of the other. They are then squeezed together in a heated press.

FAMILIAR PLASTICS

Acrylics Most familiar among the acrylic Perspex, also called Lucite. Highly tra parent and glass-like, Perspex is used for s things as watch glasses and contact len and artificial eyes. Chemically it is polyme methacrylate, abbreviated to PMMA. Acry are also familiar in fibre form, under s trade names as Acrilan and Orlon.

Bakelite, produced by Leo Baekelanc 1909, was the first synthetic plastic. I made from phenol and formaldehyde. It thermosetting plastic that is resistant to h It is therefore used for such things as sau pan handles. Its main drawback is that dark in colour. Plastics similar to it but ligh colour are melamine-formaldehyde and u formaldehyde.

Celluloid was the first man-made pla being first produced in 1869. Today its n familiar use is in table-tennis balls. It is widely used any more because it is hi flammable. Chemically, Celluloid is cellu nitrate. It is made by treating pure cellu with a mixture of sulphuric and nitric ac A little camphor is added to the cellu nitrate to make it into a workable plastic.

Cellulose acetate has replaced Celluloi many uses, for example, in the manufactur transparent sheet for use as the base photographic film. Cellulose acetate is used to make knife handles, toothbrus and toys. A modified form of it, cellu aceto-butyrate or CAB, is used for moule telephone handsets. Cellulose acetate triacetate are also widely used in fibre fc making fabrics that keep their shape well resist creasing.

Epoxy resins are plastic-like materials can readily be dissolved to make lacquers coating, for example, the inside of cans. T can also be made into powerful adhesi These come in two tubes – one contains resin, the other a hardener to make it The two are mixed together immedia before use. Such adhesives are widely u

A model of a polyethylene molecule showing its chain-like arrangement of carbon atoms (black) and hydrogen atoms (white).

Chemicals

aircraft construction and have even been ...d to stick sections of concrete bridges.

...lon, one of the most widely used plastics, probably most familiar in fibre form for ...on stockings, lingerie, shirts, dresses, ...pets, and fishing nets. The first synthetic ...e, it was produced by W. H. Carothers in ...35. In the form of a solid plastic, nylon is ...d for making gears and bearings for ...ch things as food-mixers and children's ...s. Nylon gears are 'self-lubricating' in that ...y require no oil to make them slide ...oothly and silently over one another.

...lyesters are again best known as syn-...tic fibres, under such trade names as ...cron, Terylene, and Fortrel. The fibre form ...f polyester is a thermoplastic which will ...ten on reheating. There are other forms of ...yester that will link together in the ...sence of a catalyst to form a non-...tening plastic. To give the plastic strength, ...is reinforced with glass fibre. Fibre-...ss reinforced plastic has a variety of uses ...m making fishing rods and suitcases to ...ulding car bodies and boat hulls.

...lycarbonate This has been produced ...commercial quantities only since about ...0. Polycarbonate is tough and rigid and ...a number of engineering uses. It has a ...her softening point than the other com-...n plastics and can withstand steam and ...ling water. It is thus used for utensils that ...d to be sterilized.

...lyethylene, better known in Britain under ...trade name Polythene, is one of the most ...satile of plastics. It can be made into ...tles, transparent film for packaging, pro-...tive coverings, damp-proof membranes ...house-floors, and so on. There are two ...ic kinds — soft, flexible low-density and ...d, rigid high-density polyethylene.

...lypropylene is very similar to high-...sity polyethylene in many ways. But it is ...newhat glossier and has a higher soften-...point. Like polycarbonate, it can be ...rilized by boiling and therefore has medical ...es. It is remarkably resistant to repeated bending. Thus a polypropylene box and lid can be made in one piece.

Polystyrene can be found in many guises in the home. The transparent measuring jugs used in the kitchen are made of polystyrene. White insulating ceiling tiles are made of expanded polystyrene. It is really a solid foam made by introducing liquefied gas, such as pentane, into beads of polystyrene. When heated, the beads are forced to expand by the gas inside

Polytetrafluorethylene, usually abbreviated to PTFE, is a fluorinated relative of polyethylene. Its main domestic application is for the non-stick coating on saucepans and frying pans. In industry its inertness to practically every kind of chemical even at relatively high temperatures makes it suitable for use in corrosive conditions.

Polyurethane is a plastic that is used in the form of a foam in upholstery, draught excluders, coat linings, and so on. Paints and varnishes based on polyurethanes are hard and tough.

PVC, short for polyvinyl chloride, is one of the most useful of plastics. Its many domestic uses include floor tiles, plastic handbags and shoes, guttering and drainpipes, LP records, and recording tapes. PVC can be obtained in varying degrees of hardness — guttering is very rigid, plastic shoes are very supple. The difference lies in the amount of plasticizer used in the PVC. Without the plasticizer, PVC would be too brittle to be used as a plastic.

Silicones are rather different from the other plastics here. Their long molecules are made up of chains not of carbon atoms but of silicon and oxygen atoms. And they are fluids rather than solids. In fluid form they are used to waterproof raincoats. In polishes silicones impart a hard, shiny, and waterproof finish. Silicones can be made into synthetic rubber that is very inert. It is used in surgery for it does not react with body tissues. It can also be steam-sterilized. Silicone rubbers can be used at temperatures as high as $250°C$ or as low as $-70°C$.

Chemical Elements

Actinium (Ac, atomic number 89) Rare radioactive metal found in uranium ores, such as pitchblende, resulting from the decay of uranium-235. First of the actinide series of elements.

Aluminium (Al, 13) Strong, lightweight, corrosion-resistant metal of the boron group (Group 3A), second in importance only to iron. Most abundant metal in the Earth's crust comprising 8%, but its only important ore is bauxite, containing the oxide alumina. It has an oxidation number of 3. The oxide, Al_2O_3, and hydroxide $Al(OH)_3$, are amphoteric. The hydrated mixed sulphate with potassium—potassium aluminium sulphate, or potash alum, $K_2SO_4.Al_2(SO_4)_3.24H_2O$—is typical of the series of compounds called alums.

Americium (Am, 95) Artificial transuranic radioactive element made in 1944 by bombarding plutonium with neutrons; the fourth artificial element to be made.

Antimony (Sb, 51) Hard, brittle element of weak metallic character belonging to the nitrogen family (Group 5A). It is added to some alloys (such as type metal) to harden them. In compounds it may have an oxidation number of 3, as in the poisonous gas stibnine SbH_3, or 5, as in the pentoxide Sb_2O_5. It also forms oxides with oxidation numbers of 3 and 4; the trioxide Sb_2O_3 is amphoteric; the others are weakly acidic.

Argon (Ar, 18) The most abundant of the inert, or rare, gases (Group 0), making up about 1% of the atmosphere. Obtained by fractional distillation of liquid air, it is used to fill light bulbs and Geiger counters, as an inert atmosphere in arc-welding and as a carrier gas in gas chromatography.

Arsenic (As, 33) A feebly metallic element of the nitrogen family (Group 5A), whose compounds are deadly poisons. It occurs widely in compounds, such as the sulphide realgar, As_2S_3, and the mixed sulphide with iron, mispickel, FeSAs. In its compounds it commonly has an oxidation number of 3, as in the poison gas arsine, AsH_3, and the amphoteric arsenic(III) oxide, As_2O_3. It can also have other oxidation states, as in the arsenic(V) oxide, As_2O_5, which gives rise to arsenates.

Astatine (At, 85) The rarest naturally occurring element, which has about 20 known isotopes, none of them stable. It is produced during radioactive decay of uranium, thorium and actinium. It is the heaviest member of the halogen series.

Barium (Ba, 56) A heavy alkaline-earth metal (Group 2A) which occurs in Nature as the sulphate, barytes, or heavy spar, $BaSO_4$; or as the carbonate witherite $BaCO_3$. It bursts into flame on contact with air and decomposes water, with the formation of the hydroxide, $Ba(OH)_2$, and the liberation of hydrogen.

Berkelium (Bk, 97) Artificial radioactive element obtained by bombarding americium with helium ions; the fifth transuranium element.

Beryllium (Be, 4) A lightweight alkaline-earth metal (Group 2A) with a relatively high melting point (1283°C). Strong, hard and elastic, it is used in alloys to confer these properties on other metals. It is highly transparent to X-rays. Its main source is the mineral beryl, beryllium aluminium silicate.

Bismuth (Bi, 83) A brittle metal with a reddish tinge, belonging to the nitrogen family (Group 5A). It forms low-melting point, fusible alloys with lead and iron. It resembles antimony chemically, and forms three oxides Bi_2O_3, Bi_2O_4, and Bi_2O_5. The trioxide is basic; the pentoxide is weakly acidic. The metal unites readily with sulphur and the halogens. Bismuth salts are used in soothing medicines for digestive disorders. Bismuth is rarely found native; its chief sources are the sulphide bismuthite, or bismuth glance, Bi_2S_3, and the trioxide bismite, B_2O_3.

Above: The anode blocks of an aluminium cell. Aluminium is produced by the electrolytic decomposition of alumina (aluminium oxide) dissolved in molten cryolite (sodium aluminium fluoride). It is the only economic way of producing the metal from its ores.

Boron (B, 5) A very important semi-metallic element in the same family as aluminium (Group 3A), whose most familiar compound is borax, disodium tetraborate $Na_2B_4O_7 \cdot 10H_2O$. The borax-bead test used in chemical analysis relies on the property of fused borax to dissolve metallic oxides to form metaborates, some of which are coloured —cobalt metaborate, for example, is blue. Boron can take two forms—a brown amorphous powder and a black, lustrous crystalline material. The latter is an important semiconductor. Boron is used in control rods for atomic reactors since it readily absorbs neutrons. It combines with hydrogen to form

Chemical Elements

boranes such as boroethane, or diborane (B_2H_6), whose constitution cannot readily be explained by normal theories of bonding. Boron reacts with oxygen to form the amphoteric trioxide B_2O_3; with halogens to form halides; and with nitrogen to form nitrides, including the hard-as-diamond crystalline compound, borazon, which is used as an abrasive. Boron will also combine with metals such as titanium and tungsten to form heat-resistant borides.

Bromine (Br, 35) A dark red liquid element of the halogen family (Group 7A), whose vapour is pungent and irritates the eyes and respiratory system. Bromine, which has few uses in the uncombined state, is extracted from sea water. It is highly reactive and forms useful bromides with metals and organic substances. Among the most important are light-sensitive silver bromide, which is used in photographic emulsions, and ethylene dibromide which is used, for example, in petrols to remove lead from the cylinders after combustion.

Cadmium (Cd, 48) A soft metal of the zinc group (Group 2B), which like tin crackles when deformed. It is obtained mainly as a by-product in lead and zinc smelting and refining. Its main use is for plating metals and alloys to protect them from corrosion. It is used in making control rods for nuclear reactors because it readily absorbs neutrons. It is used with nickel in long-lasting nickel-cadmium storage batteries, and in the Weston standard cell. Bright-yellow cadmium sulphide (CdS), which occurs in nature as the mineral greenockite, is an important pigment for inks and paints. Some cadmium compounds are poisonous and have been responsible for serious water pollution in some industrial regions.

Caesium (Cs, 55) A very soft alkali metal (Group 1A), which resembles sodium and potassium and reacts explosively with water. It loses electrons when struck by light and finds ready application in photoelectric devices. Vaporized caesium is used in the atomic clock

Calcium (Ca, 20) A reactive alkaline-earth metal (Group 2A) whose compounds are widespread in Nature, including the carbonate, $CaCO_3$ (*chalk* and *limestone*); the hydrated sulphate, $CaSO_4.2H_2O$ (*gypsum*); and the phosphate, $Ca_3(PO_4)_2$ (bones and teeth and the mineral *apatite*). Among its many other industrially important compounds are *quicklime*, the oxide, CaO; *slaked lime*, the hydroxide, $Ca(OH)_2$; *bleaching powder*, $Ca(ClO)_2.Ca(OH)_2.CaCl_2$, a mixture of calcium hypochlorite with basic calcium chloride; and calcium carbide, CaC_2, from which acetylene gas is made.

Californium (Cf, 98) Artificial radioactive transuranic element produced by bombarding curium with helium ions. One isotope (Cf-252) is an intense neutron emitter.

Carbon (C, 6) A typical non-metallic element (Group 4A) whose atoms have the ability to link with each other in long and complex chains to form a larger number of compounds than all the other elements put together. Study of these compounds forms the basis of organic chemistry, so called because living organisms are made up of carbon compounds. Carbon occurs in the Earth's crust as soft, flaky graphite and as hard, crystalline diamond, and in compounds with metals and oxygen as carbonates ($-CO_3$). The atmosphere contains traces of carbon dioxide (CO_2), resulting from animal respiration and from combustion of carbon-containing materials, such as wood and fossil fuels. Partial combustion of carbon yields poisonous carbon monoxide CO, which is a powerful reducing agent. Carbon dioxide forms the weak carbonic acid in water, whose salts are the carbonates. With sulphur, carbon forms the poisonous and flammable disulphide (CS_2), which is a volatile liquid; large amounts of carbon disulphide are used in the manufacture of rayon and Cellophane. With the halogens, carbon forms compounds such as carbon tetrachloride CCl_4, a non-flammable liquid used for dry cleaning and in some fire extinguishers. With hydrogen, carbon forms the so-called hydrocarbons, such as methane (CH_4) and ethane (C_2H_6), which form the main constituents in natural gas and petroleum. The radioactive isotope

Chemical Elements

carbon-14 is formed in Nature, and provides a means of dating archaeological specimens.

Cerium (Ce, 58) One of the rare-earth metals (transition Group 3B), which is quite plentiful in the Earth's crust, occurring in the minerals cerite and monazite, for example. The cerium alloy, misch metal, is used in making lighter flints. The yellow ceric sulphate, $Ce(SO_4)_2$, is a strong oxidizing agent used in volumetric analysis.

Chlorine (Cl, 17) Industrially the most important of the halogens (Group 7A), chlorine is a poisonous yellowish-green gas that irritates the eyes and respiratory system. Chlorine combines with both metals and non-metals. It combines with hydrogen to form hydrogen chloride, which is strongly acidic and gives rise to a wide range of metal salts, such as sodium chloride (NaCl). The gas is prepared by the electrolysis of brine. It and its compounds are excellent bleaching agents.

Chromium (Cr, 24) A hard, brittle transition metal (Group 6B) which takes a high polish and finds widespread application in chromium plating. It confers strength and corrosion resistance to alloys with steel. It was called chromium (colour) because of the colour exhibited by many of its compounds, such as yellow lead chromate, $PbCrO_4$, or green chrome oxide, Cr_2O_3. The red of the ruby and the green of the emerald are also due to chromium compounds. The chromates and dichromates are strong oxidizing agents. Potassium dichromate, $K_2Cr_2O_7$, is used in volumetric analysis. The mixed oxide with iron, *chromite*, $FeCr_2O_4$, is the chief ore.

Cobalt (Co, 27) A transition metal (Group 8B) used in high-temperature and magnetic alloys, such as *alnico* (with aluminium and nickel). It resembles iron in its chemistry, but physically it is heavier, stronger, and harder than iron. It forms more complex ions than any other metal except platinum. The radioactive isotope cobalt-60 is widely used in radiation therapy.

Copper (Cu, 29) An attractive, corrosion resistant, reddish-brown transition metal (Group 1B) widely used on account of its high electrical and thermal conductivity and its ability to form strong, corrosion-resistant alloys, such as bronze (with tin), brass (with zinc) and cupronickel (with nickel). Found native and in sulphides, oxides and carbonates, copper is widely distributed.

Curium (Cm, 96) Artificial, highly radioactive transuranium element made by bombarding plutonium with accelerated helium ions.

Dysprosium (Dy, 66) One of the rare-earth metals (transition Group 3B), whose compounds are sometimes used as catalysts. It has interesting magnetic properties. Below $-168°C$ it becomes ferromagnetic, like iron; at really low temperatures it becomes superconductive.

Einsteinium (Es, 99) Artificial, short-lived, radioactive transuranium element made by irradiating uranium-238 with neutrons. It was first discovered in the debris from the first thermonuclear (H-bomb) explosion in 1952, and was synthesized two years later.

Erbium (Er, 68) One of the rare-earth elements (transition Group 3B), many of whose compounds are pink, including the rose pink oxide Er_2O_3. At very low temperatures it is ferromagnetic and superconductive.

Europium (Eu, 63) The lightest and softest of the rare-earth elements (transition Group 3B), it is also one of the least abundant. Many of its salts are coloured. It readily absorbs neutrons and has potential applications in the nuclear energy field.

Fermium (Fm, 100) Artificial, radioactive transuranium element, obtained by bombarding uranium-238 with neutrons. Like einsteinium, it was first discovered in the debris from the first H-bomb, and subsequently synthesized.

Fluorine (F, 9) A greenish-yellow choking gas which is the first member of the halogen family (Group 7A) and the most reactive of all the elements. It combines with every other element except the rare gases helium, neon, and argon to form fluorides. It occurs widely in mineral compounds, including *fluorite* (*fluorspar*) CaF_2; *apatite*, $CaF.3Ca_3(PO_4)_2$; and *cryolite*, sodium aluminium fluoride, Na_3AlF_6. Boron and antimony fluorides are

Chemical Elements

important catalysts; sodium fluoride is added to drinking water to help reduce tooth decay (fluoridation); uranium hexafluoride is used in separating uranium isotopes. Fluorine reacts with hydrocarbons to form fluorocarbons, including the 'non-stick' plastic, Teflon, which is polytetrafluorethylene $(CF_2 CF_2)_n$; and Freon, dichlorodifluoromethane (Cl_2CF_2), a refrigerant.

Francium (Fr, 87) Rare radioactive element which is the heaviest of the alkali metals (Group 1A). It occurs transiently during radioactive decay of actinium. Only a few grams are present in the Earth's crust at any moment. Its properties resemble those of caesium.

Gadolinium (Gd, 64) A typical rare-earth metal (Group 3B) found with many others in the mineral *monazite*. Its salts and solutions are colourless. It becomes ferromagnetic at temperatures of 17°C and below, and at very low temperatures becomes superconductive. It absorbs neutrons more readily than any other element.

Gallium (Ga, 31) A soft metal of the boron group (Group 3A) which melts at just above room temperature (at 29·8°C). Once called eka-aluminium, it has somewhat similar chemical properties to aluminium. Its compounds occur sparsely in Nature, and it is extracted as a by-product from ores such as zinc blende and bauxite. Like aluminium it is

Panning for gold in southern Africa. In hand panning the miner swirls round a handful of gravel from a stream bed with water. Light earthy material is washed away, and any gold present remains in the pan.

Chemical Elements

amphoteric, reacting with both acids to form salts and alkalis to form gallates. With phosphorus, arsenic and antimony, gallium forms compounds that are semiconductors—gallium arsenide in particular is widely used.

Germanium (Ge, 32) A rare semi-metallic element (Group 4A) which, like its sister element silicon, has valuable semiconductor properties. Among its important compounds are the oxide, GeO_2, and the tetrachloride, $GeCl_4$. Zinc germanate, Zn_2GeO_4, is used as a phosphor in fluorescent lamps. Germanium is extracted from certain ores, such as *germanite*, and can sometimes be recovered from flue dust—for many coals contain tiny amounts of the element.

Gold (Au, 79) Bright yellow, unreactive precious metal (transition Group 1B), which conducts heat and electricity superbly, and which is the most malleable and ductile of all metals. It occurs native and is widely mined. One of its important industrial uses is for plating contacts in printed circuits and semiconductors. Of the acids only aqua regia—a mixture of concentrated hydrochloric and nitric acids—will dissolve it.

Hafnium (Hf, 72) Transition metal (Group 4B) closely resembling zirconium in chemical properties. It is hard, malleable and ductile, and has excellent corrosion resistance. One of its main uses is in control rods for nuclear reactors. It occurs naturally as the oxide in zirconium minerals such as zircon and cyrtolite.

Hahnium (Ha, 105) A short-lived radioactive transuranium element, whose most stable isotope has a half-life of only a few seconds.

Helium (He, 2) The lightest element after hydrogen and first member of the inert gases (Group 0). It has the lowest boiling point of any element ($-269°C$), and in liquid form displays the phenomenon of superfluidity. It was discovered first in the Sun's atmosphere. After hydrogen, it is probably the most common element in the universe. It is synthesized from hydrogen during the nuclear fusion processes which produce stellar energy.

Gold bars stacked in the vaults of the Bank of England. Each one has a mark which identifies the company that cast it. Gold still forms the basis of most world monetary systems, even though it has long since ceased to be used for coinage.

Holmium (Ho, 67) One of the least abundant of the rare earths (transition Group 3B), which forms yellowish-brown salts. It becomes ferromagnetic at very low temperatures (below $-253°C$).

Hydrogen (H, 1) The simplest and lightest of all elements, and the most abundant element in the universe. Chemically it bears certain resemblances to both the alkali metals (Group 1) and the halogens (Group 7). It forms hydrides with weak metals, as in tin hydride (SnH_4); and with non-metals, as in hydrogen fluoride and chloride (HF, HCl); nitrogen hydride, or ammonia (NH_3); and carbon hydride, or methane (CH_4). With carbon it forms an enormous range of organic compounds.

Indium (In, 49) Rare metal of the boron family (Group 3A), which is soft and plastic

Chemical Elements

and like tin emits a 'cry' when bent. It was named after the indigo blue colour displayed by its compounds in the flame test. Its oxide, In_2O_3, is yellow. It is used in bearing alloys to improve lubricant wetting characteristics and corrosion resistance. Its most important use is in the manufacture of semiconductor devices.

Iodine (I, 53) Soft, dark grey crystalline solid belonging to the halogen family (Group 7A). It sublimes at room temperature to give a deep violet irritating vapour. It occurs naturally in salt deposits and brine, as iodides, such as sodium iodide, NaI; and iodates, such as calcium iodate, $Ca(IO_3)_2$. It can also be extracted from seaweeds. Tincture of iodine—iodine dissolved in potassium iodide, KI, water and alcohol—is an excellent bactericide. Iodine combines with most metals and non-metals. Hydrogen iodide (HI) in solution is a strong acid, giving rise to salts—iodides. Iodic acid (HIO_3) gives rise to the iodates.

Iridium (Ir, 77) Rare brittle metal of the platinum group (transition Group 8B) which is one of the hardest and densest pure metals known (relative density 22). It finds its greatest use in alloys with platinum, and gives that metal greater chemical resistance. It is obtained in alloys with other platinum metals as a by-product of nickel and copper refining.

Iron (Fe, 26) A weak, corrosion-prone transition metal (Group 8B) that can be transformed by the addition of small amounts of alloying materials, particularly carbon, into our most important metal, steel. It is the fourth most abundant element and the second most abundant metal (after aluminium) in the Earth's crust, where it occurs as oxides, such as haematite (Fe_2O_3) and magnetite (Fe_3O_4); sulphides, such as pyrites (FeS); and carbonates, such as siderite ($FeCO_3$). One of iron's most distinctive properties is its magnetism. Like many transition metals, iron forms complex ions, such as hexacyanoferrates (ferricyanides), as in the potassium salt, $K_3Fe(CN)_6$.

Krypton (Kr, 36) One of the inert gases (Group 0); found in small traces (about one part per million) in the atmosphere. It is used to fill some electric tubes and fluorescent lamps.

Lanthanum (La, 57) The first member of the lanthanide series of rare-earth metals (Group 3B)—fifteen metals with very similar chemical and physical properties. Lanthanum is found with other rare earths in monazite and other minerals. The alloy, misch metal, which is used for lighter flints, contains 25% lanthanum.

Lawrencium (Lw, 103) The eleventh transuranium element, named after the inventor of the cyclotron, Ernest Lawrence. It is made by bombarding californium with accelerated boron ions.

Lead (Pb, 82) A soft, heavy metal related to tin (Group 4A). Easy to shape and resistant to corrosion, it has been used for making water pipes for thousands of years. Today its greatest use is in car batteries, though it is widely used elsewhere in numerous alloys. Its soluble compounds are poisonous.

Lithium (Li, 3) The first of the alkali-metal group (Group 1A) and the lightest of all solid elements (relative density 0·53). A soft, silvery white metal of low melting point (180°C), it is used in some lightweight alloys. Like all the alkali metals, it decomposes water, liberating hydrogen. (See page 205.)

Lutetium (Lu, 71) Last member of the rare-earth metals (Group 3B), of which it is the hardest and densest and has the highest melting point. It is of little use.

Magnesium (Mg, 12) A light alkaline-earth metal (Group 2A), it is widely used in the aerospace industries in alloys with aluminium and other metals. It burns brilliantly in air to form pure white magnesium oxide. It occurs widely, for example, as the carbonate magnesite and the double carbonate with calcium, dolomite.

Manganese (Mn, 25) A hard transition metal (Group 7B), whose main use is in steelmaking. Manganese absorbs impurities from the steel and strengthens it. High manganese steel has exceptional hardness and wear resistance. Important manganese compounds include the oxide, which is a good catalyst, and potassium permanganate, an excellent oxidizing agent. Manganese has oxidation states of 2+ in salts, 6+ in manganates and 7+ in permanganates.

Mendelevium (Md, 101) Artificial radioactive element of the actinide series, obtained by bombarding einsteinium with helium ions. Relatively short-lived, its most stable isotope has a half-life of only two months.

Mercury (Hg, 80) The only metal that is liquid at room temperature, mercury is related to zinc and cadmium (Group 2B). Often called quicksilver, mercury is used in barometers and thermometers; and as vapour in discharge tubes and vacuum pumps (mercury diffusion pumps). It forms alloys called amalgams with most metals (but not iron); silver and gold amalgams are used to fill teeth. It forms mercury (I) compounds (e.g. calomel, Hg_2Cl_2) and mercury (II) (e.g. $HgCl_2$). Both chlorides have medical uses. The main source of mercury is cinnabar (HgS).

Molybdenum (Mo, 42) A transition metal (Group 6B) which is alloyed with steel to impart high-temperature strength. It has a high melting point (2620°C) and retains its hardness and strength at high temperatures. One of its most useful compounds is the disulphide MoS_2, which has a layered structure (like graphite), and is a lubricant.

Neodymium (Nd, 60) The third most plentiful of the rare-earth metals (Group 3B). Found in the mineral monazite, it is a constituent of the alloy, misch metal (for lighter flints), and is used to colour ceramic glazes and glasses.

Neon (Ne, 10) One of the most useful of the rare gases (Group 0), neon is used in discharge tubes. Neon tubes emit a brilliant orange-red light. Neon is obtained by distilling liquid air.

Neptunium (Np, 93) An artificial radioactive element following uranium (which it resembles) in the Periodic Table. It is obtained by bombarding uranium with slow neutrons. Its most stable isotope has a half-life of more than 2 million years.

Nickel (Ni, 28) A tough, hard silvery white transition metal (Group 8), which has excellent corrosion resistance. Its alloy with copper, cupronickel, is used for our 'silver' coinage. It is widely used for plating, in stainless steels, and as a catalyst. One of its most interesting compounds is the volatile carbonyl, $Ni(CO)_4$. Being ferromagnetic, nickel is a major ingredient of magnet alloys.

Niobium (Nb, 41) A soft, ductile metal closely related to tantalum (Group 5B) with which it occurs in ores such as columbite. (It was once called columbium.) A major use is in superconductors.

Nitrogen (N, 7) A colourless, tasteless, and odourless gas (Group 5A), nitrogen is the major constituent of air (78% by volume). It is an important constituent of the proteins in living matter. It occurs in Nature as nitrates, such as Chile saltpetre, $NaNO_3$; as ammonium compounds, such as NH_4Cl. It forms a whole series of oxides, including laughing gas, N_2O.

Nobelium (No, 102) An artificial radioactive element made by bombarding curium with carbon ions. The most stable isotope has a 3-month half-life.

Osmium (Os, 76) A hard, brittle metal that has a greater density (relative density 22·5) than any other. One of the platinum metals (Group 8), it has an exceptionally high melting point (3000°C). It is used chiefly as a hardener in platinum alloys.

Oxygen (O, 8) The life-giving gas in air of which it makes up 21% by volume. It is the most abundant element (Group 6A) in the Earth's crust, of which it makes up 46·6%. It combines with most other elements to form oxides, such as water (H_2O), carbon dioxide (CO_2), silicon dioxide (SiO_2), iron oxides (Fe_2O_3, Fe_3O_4) and aluminium oxide, bauxite (Al_2O_3). Liquid oxygen, obtained by distilling liquid air, is used as a rocket propellant and explosive. (See page 216.)

Palladium (Pd, 46) A ductile, corrosion-resistant transition metal (Group 8) similar to platinum. It is used as a catalyst, and to make jewellery and electrical contacts. (See

Phosphorus (P, 15) A waxy, solid element of the nitrogen family (Group 5A) which glows in the dark. It bursts into flame on contact with air. An essential ingredient in living matter, phosphorus occurs in teeth

Chemical Elements

and bones as calcium phosphate. There are four allotropes of phosphorus—white, red, violet and black, in order of reactivity.

Platinum (Pt, 78) A heavy, soft, ductile and corrosion-resistant transition metal (Group 8) with high melting point (1769°C). Its main use is in jewellery and as a catalyst. It has a high electrical resistance, which varies with temperature. This property is used in the platinum resistance thermometer.

Plutonium (Pu, 94) An artificial radioactive element made by bombarding uranium-238 with neutrons. It is the most important of the transuranic elements because its isotope Pu-239 is fissile, and can therefore be used as fuel in nuclear reactors (see page 30). Plutonium wastes from nuclear reactors pose a serious pollution threat because they are deadly poisons and they have a long half-life (24,000 years).

Polonium (Po, 84) A naturally-occurring radioactive metal (Group 6A) found in traces in pitchblende and other uranium minerals. The most common isotope has a half-life of 138 days.

Potassium (K, 19) One of the soft, reactive alkali metals (Group 1A) closely resembling sodium in chemical properties. It was the first metal to be isolated by electrolysis (by Humphry Davy, 1807).

Praseodymium (Pr, 59) A yellowish rare-earth metal (Group 3B) which forms greenish salts. It is used to colour ceramics, in lighter flints, and in some lightweight alloys.

Promethium (Pm, 61) A radioactive rare-earth element (Group 3B) produced during nuclear fission of uranium and by bombarding neodymium with neutrons. It has typical rare-earth properties, and most of its compounds are pink.

Protactinium (Pa, 91) A very rare naturally-occurring radioactive metal preceding uranium in the Periodic Table. Traces of it occur in all uranium ores. The most stable of its 12 isotopes has a half-life of 33,000 years.

Radium (Ra, 88) A heavy, rare, radioactive alkaline-earth metal (Group 2A), first isolated in 1910 by Madame Curie. It results from the decay of uranium, and is found in uranium ores such as pitchblende. The penetrating rays given off by radium are used in cancer therapy. The most stable radium isotope has a half-life of 1600 years.

Radon (Rn, 86) The heaviest of the inert rare gases (Group 0), which is formed when radium decays. Like radium, it is radioactive. Traces of radon appear in the air, but in hardly detectable concentrations. Even the most stable of its 17-odd isotopes has a half-life of only 4 days.

Rhenium (Re, 75) A very rare and expensive transition metal (Group 7B) with exceptional hardness and resistance to wear and corrosion. It also has a very high density (relative density 21) and its melting point (3180°C) is the highest of all metals except tungsten. It is found naturally in the molybdenum ore, molybdenite, and in other sulphide ores.

Rhodium (Rh, 45) One of the precious platinum family of transition metals (Group 8). Ductile and corrosion-resistant, it is often plated on to silver to prevent tarnishing. It is added to platinum as a hardener.

Rubidium (Rb, 37) A soft alkali metal (Group 1A) of similar properties to sodium, which will burn in air and decompose water. Two isotopes occur in Nature. One, Ru-87, is radioactive, with a 50,000 million year half-life; it decays to strontium-87. One form of radiometric dating uses the rubidium-strontium decay.

Ruthenium (Ru, 44) A hard, brittle transition metal (Group 8) of high melting point related to platinum. It is alloyed with platinum to harden it.

Rutherfordium (Ru, 104) The name suggested for the transuranium element 104 by American physicists. Russian physicists, who claim to have made it in 1964, five years before the Americans, call it kurchatovium.

Right: Much of the colour in firework displays comes from compounds of the alkaline-earth metals strontium and barium. Strontium is responsible for the vivid crimsons, barium for the greens.

Chemical Elements

Samarium (Sm, 62) A rare-earth metal (Group 3B), obtained from the mineral monazite. It has uses in ceramics and electronics and as a catalyst in the chemical industry. Its salts are red to yellow.

Scandium (Sc, 21) The first member of the first transition series of metals (Group 3B). A rare-earth metal, it was first identified in Scandinavian ores containing other rare earths. Traces of scandium are also found in tin and tungsten ores. Many stars, including the Sun, contain appreciable amounts of scandium.

Selenium (Se, 34) A metalloid belonging to the oxygen family of elements (Group 6A) and related chemically to sulphur and tellurium. It often occurs as the selenide with the sulphide ores of lead, silver, and copper. It can be obtained from the flue dust when these ores are roasted, or from the slime at the anode in electrolytic refining. Of its many different forms, the most important is the metallic. The metallic form is used in photoelectric devices because it emits electrons when light strikes it.

Silicon (Si, 14) The most abundant element (27·7%) in the Earth's crust after oxygen, silicon is a non-metal closely related to carbon (Group 4A). It occurs almost everywhere in the crust as silicates, such as feldspar, $KAlSi_3O_8$; or as the dioxide, silica or crystalline quartz, SiO_2. Pure silicon is a hard, metallic-looking solid with the crystal structure of diamond. Adulterated with traces of other elements, silicon becomes a semiconductor. Minute silicon chips form the basis of the large-scale integrated (LSI) circuits used in many solid-state electronic devices, such as pocket calculators. Silicon is also used to make the solar cells which provide spacecraft with electricity.

Silver (Ag, 47) A precious metal related to

Chemical Elements

copper and gold (Group 1B). Its beautiful whiteness, the ease with which it can be shaped, and its resistance to most corrosion led to its use in jewellery, for coins, and for expensive tableware. The scientist prizes it because it conducts heat and electricity better than any other element. Many of its salts are light-sensitive.

Sodium (Na, 11) A soft, alkali metal (Group 1A) which is the sixth most abundant (2·80%) element in the Earth's crust. One of the most widespread sodium compounds is the chloride, common salt (NaCl). One of the most reactive of all elements, sodium decomposes water to yield sodium hydroxide, or caustic soda (NaOH), which is a strong alkali.

Strontium (Sr, 38) An alkaline-earth metal (Group 2A) whose compounds colour a flame a characteristic crimson. It occurs in minerals such as the carbonate strontianite ($SrCO_3$) and the sulphate celestite ($SrSO_4$). It closely resembles calcium (see page 207).

Sulphur (S, 16) A non-metallic element related to oxygen (Group 6A), sulphur is one of the few elements that can be found native in the Earth's crust. It also occurs in the form of metal sulphides—those of iron (FeS), lead (PbS), and zinc (ZnS), for example; and as sulphates, such as gypsum ($CaSO_4.2H_2O$). The most important chemical made from sulphur is sulphuric acid, H_2SO_4 —the 'life-blood' of industry. (See page 230.)

Tantalum (Ta, 73) A very hard, acid-resisting heavy metal with an extremely high melting point (3000°C). It is a transition metal (Group 5B) closely related to niobium. It occurs with that metal in the columbite-tantalite series of minerals. Its main uses are in capacitors and in corrosion-resistant chemical equipment.

Technetium (Tc, 43) A radioactive transition element (Group 7B), which was the first element to be produced synthetically by nuclear bombardment. Large quantities of technetium are now produced in nuclear reactors as products of nuclear fission.

Tellurium (Te, 52) A metalloid in the oxygen group (Group 6A) which closely resembles selenium. It is obtained as a by-product of lead, gold and copper refining, occurring in the ores as a telluride. It is occasionally used in alloys, and in semiconductors—such as bismuth telluride.

Terbium (Tb, 65) One of the rarest of the rare-earth metals (Group 3B), it is sometimes used in lasers and semiconductors.

Thallium (Tl, 81) A soft metal (Group 3A) whose physical properties closely resemble those of lead, which follows it in the Periodic Table. Its compounds are very poisonous.

Thorium (Th, 90) A naturally occurring radioactive metal of the actinide series (Group 3B). Occurring in such minerals as monazite and thorite, thorium is three times more abundant than uranium. In breeder reactors the isotope Th-232 will change into uranium-233, which is fissile. The dioxide, ThO_2, is an important industrial refractory.

Thulium (Tm, 69) A fairly rare rare-earth metal (Group 3B). Its short-lived radioactive isotope, Tm-170, is sometimes used as a portable X-ray source since it emits soft gamma-radiation resembling X-rays.

Tin (Sn, 50) A soft, ductile corrosion-resistant metal of the carbon family (Group 4A). It is widely used as plating on mild steel, as tinplate, and in many alloys, including solder, bronze, type metal and pewter. It occurs in two allotropic forms—white and grey. The familiar form is white which, when pure, changes into the powdery grey form at low temperatures. Impure commercial tin does not usually undergo such a change.

Titanium (Ti, 22) A strong, light transition metal (Group 4B) which has excellent corrosion resistance. It is widely used in the aerospace industries. It is also used in surgical aids. The ninth most abundant element, it is found in the minerals, ilmenite and rutile. Rutile—titanium dioxide—is an important white paint pigment. (See page 204.)

Tungsten (W, 74) Also called wolfram; a very strong, brittle, transition metal (Group 6B) which has the highest melting point among metals (3380°C). It is incorporated in steel to increase high-temperature strength. Tungsten wire is used for electric-light bulb filaments. Tungsten carbide is used making very hard, tough tools and dies.

Uranium (U, 92) A naturally occurring

Chemical Elements

Sodium-vapour street lamps emit the vivid orange-yellow glow characteristic of sodium compounds. They are excellent for lighting, but distort the colours of skin, fabrics, and so on.

radioactive heavy metal of the actinide series (Group 3B). Natural uranium is made up mainly of the isotope U-238, but with traces of U-235 and U-234. U-235 has the property of undergoing fission, which makes possible commercial nuclear power.

Vanadium (V, 23) A rare, soft transition metal (Group 5B) whose main use is in alloy steels. It improves the steel's hardness, strength and shock resistance. It was named after the Scandinavian goddess of beauty, for solutions of its compounds display beautiful colours. Vanadium pentoxide is an important catalyst, used, for example, in sulphuric acid manufacture by the contact process.

Xenon (Xe, 54) A very rare inert gas (Group 0), more than four times as heavy as air. It is used in high-intensity flash lamps. It was the first inert gas to be made to combine chemically with other elements. The first compound formed was the red solid xenon hexafluoroplatinate.

Ytterbium (Yb, 70) A soft rare-earth metal (Group 3B) which forms pale green or white salts. It is scarce, expensive and little used.

Yttrium (Y, 39) An abundant rare-earth metal (Group 3B) found in such rare-earth ores as xenotime and gadolinite. Yttrium compounds are used in ceramics, in lasers, and in red phosphors for colour television. A common phosphor is yttrium orthovanadate, activated by europium.

Zinc (Zn, 30) A common metal related to cadmium and mercury (Group 2B). Mixed with copper, it forms the important alloy brass; it is coated on steel to prevent rusting (galvanizing). It is the negative pole in dry batteries. Its most important commercial ore is zinc blende, or sphalerite, ZnS.

Zirconium (Zr, 40) A soft, ductile, corrosion-resistant transition metal (Group 4B) related to titanium. Its major use is for cladding fuel rods in nuclear reactors. It has low neutron absorption, and remains strong at high temperatures. Its oxide, zirconia, is an excellent refractory.

Iron and Steel

Of all the metallic elements, iron is the most important. It is very abundant, easy to extract from its ores, and can readily be made into our primary constructional material, *steel*. No metal approaches it in terms of tonnages used. Current world production of steel is about 600 million tonnes a year. This compares with only about 10 million tonnes for aluminium, the next most widely used metal.

Metallic iron itself is a greyish-white metal that is highly ductile and malleable, and therefore easy to shape. It is magnetic; most magnets contain iron, along with other magnetic elements like nickel and cobalt. By itself, iron is neither hard nor particularly strong. But with the addition of a little carbon it becomes steel and herein lies its great value. The addition of as little as one-tenth of one per cent of carbon to iron transforms it into a metal of high tensile strength, which can be made even tougher and harder by heat treatment. Other metallic and non-metallic elements can be incorporated in steels to give them enhanced properties, for example, corrosion resistance and even greater strength and toughness.

Iron has been widely used for at least 3000 years. When people discovered the advantages of iron over bronze for making weapons and implements a new era dawned—the Iron Age. It began in about 1200 BC in the near East and south-eastern Europe, and later elsewhere. Steel of a kind was made by the Romans, but

Right: The famous iron-ore mining region known as Iron Knob, in South Australia.

Iron and Steel

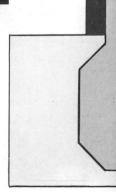

Oil Burner

it did not become available on a large scale until 1856, when Sir Henry Bessemer developed his process of steelmaking.

Iron Making

Iron is, after aluminium, the most abundant metallic element in the Earth's crust, comprising one-twentieth of it. In the Earth as a whole it is undoubtedly *the* most abundant element, for the central core of the planet is believed to be mainly iron, probably in a

Iron and Steel

Left: In a giant steel-mill, molten steel is tapped from the furnace and poured into ingot moulds.

Layout of an open-hearth steel-making furnace. The charge is melted in a shallow hearth which is open to the furnace flames. The incoming air is preheated by the hot brickwork of one of the regenerators. The brickwork in the other regenerator is being heated by the exiting furnace gases.

fluid state. Being a relatively reactive element iron is seldom found in a native state, though it can sometimes be found alloyed with nickel. Some iron-nickel alloys have been found that have reached the Earth as meteorites.

Iron compounds are widespread in the Earth's crust, and they have become concentrated in workable mineral deposits in many parts of the world. Among the many iron-bearing minerals are the oxides *haematite* (Fe_2O_3) and *magnetite* (Fe_3O_4); the hydrated oxide, *limonite* ($Fe_2O_3 . 3H_2O$); the carbonate, *siderite* ($FeCO_3$); and the sulphide, *iron pyrites* (FeS). The latter is used as a source of sulphur, not iron, but the others are all important iron ores. Haematite, often called *kidney iron ore* because of the way it sometimes crystallizes, is deep red when pure. Magnetite is black and

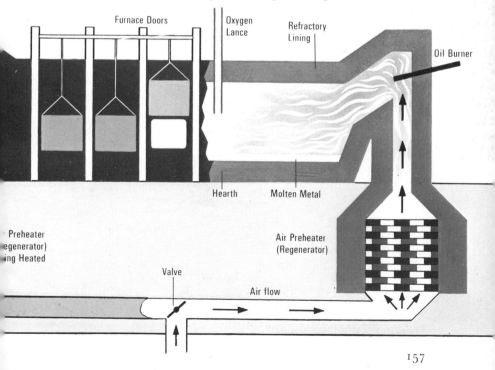

Iron and Steel

is magnetic. Limonite is brown and is often called *bog-iron ore*. Russia is the world's largest producer of iron ore, with an annual output of nearly 200 million tonnes. This is double the output of the United States, the next largest producer. Most iron ore is obtained by open cast surface mining, though some is worked in deep-level underground mines.

Because they come from the ground, most iron ores are associated with gangue, worthless earthy material which must subsequently be removed. This may be done by washing; or it may be done magnetically. Often the ore is roasted to drive off water and, in the case of carbonate ores, carbon dioxide. The ore is usually crushed during preparation, resulting in a fine dust. To prevent trouble in later operations the crushed roasted ore is heated strongly, or sintered. This treatment

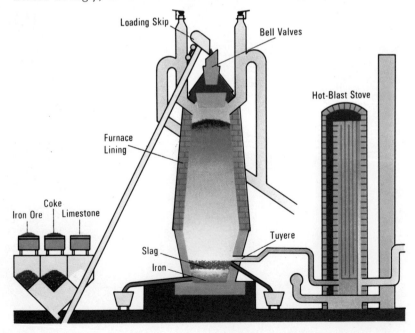

A blast furnace. Iron ore, coke and limestone are charged into the furnace through so-called bell valves, which allow material to enter without letting the furnace gases escape. Hot air from hot-blast stoves is blasted through the tuyères and causes the coke to burn fiercely, raising the temperature in the lower part of the furnace to over 1500 °C. The iron ore is reduced to iron, which runs to the bottom of the furnace. Impurities combine with the limestone to form a slag, which floats on top of the iron. Iron and slag are tapped periodically.

Iron and Steel

Above: A steel casting emerging from an annealing furnace. Annealing is one form of heat treatment, which relieves stresses in the metal.

makes it coalesce into larger lumps. Sometimes fine ore dust is mixed with clay, formed into shapes, and baked, a process called *pelletization*.

Iron is produced from the ore in *blast furnaces*. These are tall structures, 30 metres or more high and about 9 metres in diameter at the base. The essential reaction that takes place in a blast furnace is one of reduction. The iron oxide is reduced to metallic iron by a series of reactions involving carbon monoxide and carbon, which is added to the furnace in the form of coke. The coke, burning fiercely in the air blasted into the furnace, also provides the heat to melt the charge. Limestone is introduced into the furnace to act as a flux. It combines with impurities associated with the iron ore, and enables them to melt at a lower temperature than they would otherwise, to form a molten slag. This floats on top of the molten iron. The furnace is tapped periodically to allow the molten metal and slag to be removed.

Pig, Cast and Wrought Iron
The iron tapped from the blast furnace is often carried in huge travelling ladles to an adjacent steel-making plant. Or it may be poured into moulds called *pigs*, from which it gets its name, *pig iron*. Pig iron is by no means pure; it contains up to four per cent carbon and smaller amounts of other elements, which may include manganese, phosphorus, silicon and sulphur.

With slight refining, however, it becomes suitable for making iron castings. At the foundry pig iron is remelted in a *cupola*, which is a kind of miniature blast furnace. Iron scrap and other metals may be added to the pig iron to give improved properties. The iron produced, known as *cast iron*, is very fluid when molten and is thus easy to cast. The commonest cast iron, called *grey cast iron*, is strong, easy to machine and resistant to shock; but it is brittle.

Another form of iron sometimes produced from pig iron is *wrought iron*. It consists of almost pure iron mixed with threads of slag. It is made by remelting pig iron in a furnace with iron ore. Impurities in the iron combine with the iron ore to form a slag. The temperature is raised, and the carbon in the iron reacts

Iron and Steel

The piping and reactor vessels of this sulphuric-acid making plant are fabricated in stainless steel, which resists corrosion strongly.

with the iron ore. The resulting pasty iron-slag mix is then hammered or pressed to squeeze out excess slag. Although little used today, wrought iron nevertheless has excellent properties. It is tough and is highly ductile and malleable; it is resistant to shock and more resistant to corrosion than ordinary iron and steel.

Steel was produced cheaply and in large quantities for the first time in 1856 when Henry Bessemer invented the steel-refining process that now bears his name. In the process, air was blown through pig iron to

Iron and Steel

Above: Steel rod cooling after emerging from bar and rod coiling machines. Vast quantities of steel rod are used, for example, in the construction industry to reinforce concrete.

remove impurities. Today most steel refining is done in an open-hearth furnace. Pig iron, iron ore, steel scrap, and limestone are heated on a shallow hearth by the furnace flames. The flames burn out the excess carbon, while other impurities pass into the lime slag.

It takes some 10 hours to convert 350 tonnes of iron into steel in an open-hearth furnace. Gradually it is being superseded by faster processes. The basic-oxygen process can convert 350 tonnes of iron into steel in only 40 minutes. An electric-arc furnace can produce 150 tonnes of pure steel in about four hours.

Iron and Steel

The basic-oxygen furnace is shaped somewhat like the drum of a concrete mixer. It is filled with steel scrap and molten pig iron, on to which pure oxygen is blown at very high speed. Carbon and other impurities are burnt out. Lime is added during the 'blow' and combines with some impurities to form a slag. After 40 minutes the converter is tilted on its side, and the steel is poured off.

The electric-arc furnace is fed a charge of steel scrap rather than hot metal. Three carbon electrodes are lowered into the furnace, and an electric arc is struck between the electrodes and the charge. This gives out great heat, which causes the charge to melt. Lime, fluorspar, and iron ore are added to the molten metal to absorb impurities. The electric-arc furnace can produce steel of higher quality than other processes because the metal is uncontaminated by combustion products.

Above left: An electrical resistance welding set-up, showing the disc-like electrodes which not only carry the current but also apply the pressure.

Automatic rolling mill for steel sheet. A red-hot, thick steel slab enters the mill at one end and is progressively reduced in thickness as it passes between a series of rollers.

Copper

Like its close relatives gold and silver, copper has been known to Man from early times, probably as long as 10,000 years ago. All three metals can be found native in the ground, and undoubtedly the early civilizations used native copper to make utensils and ornaments.

Copper is still valued today for its attractive appearance, ease of shaping, and resistance to corrosion. It can be shaped by rolling, pressing, forging and drawing. It is so ductile that it can be drawn into wire 0·005 centimetre in diameter; so malleable that it can be hammered into strips 0·0025 centimetre thick. Joining copper parts together also presents no problem, for the metal is easy to solder, braze and weld. Most of copper's many alloys—with tin (*bronze*), zinc (*brass*), nickel (*cupronickel*)—share copper's desirable qualities to a greater or lesser extent.

One thing they do not share, however, is copper's most notable quality—its very high electrical conductivity. Only silver is superior to copper in this respect, but its cost and scarcity precludes its use on a large scale.

Refining

The processing of copper ores takes place in several stages; a number of methods are used, depending on the type of ore being handled. The extraction of copper and nickel from the mixed-sulphide Sudbury ore is outlined on page 167. Normally sulphide ores are separated from the gangue, or earthy material they are associated with, by selective *flotation*. The ore is finely ground and mixed

A Benin bronze. The kingdom of Benin was an advanced civilization of the region of West Africa which is now Nigeria. It was at its peak between about 1400 and 1600, although it continued until the British conquered it in the 1890s. Benin bronzes are among the best art Africa has ever produced.

with water and a frothing agent; the mineral particles are preferentially attracted by the bubbles and floated away. The copper concentrate from this so-called mineral-dressing operation is roasted to convert iron into iron oxide; to expel some of the sulphur; and to drive off volatile matter containing, for example, antimony and arsenic. Next the roasted ore is transferred to a reverberatory furnace, where the iron oxide combines with silica in the ore to form a slag, leaving fused sulphides of iron and copper—a mixture known as *matte*. Air is then blasted through the molten matte in a converter to burn away the sulphur. Silica is added to extract further impurities and form a slag. The resulting crude, 'blister' copper is usually then refined to high purity by electrolysis, using an electrolyte of acidified copper sulphate.

Oxide ores of copper are usually processed by *leaching*, or treatment with a substance that dissolves the mineral but not the associated impurities.

Chemical Properties

Copper is a typical transition metal in Group 1B of the Periodic Table, which also includes silver and gold. It has many characteristics of these metals, but is less corrosion resistant and more reactive generally. In its compounds it may be monovalent, copper(I), or divalent, copper(II). The divalent compounds are the commonest and most stable.

In air, copper slowly tarnishes, but if **heated to red heat, black scales of copper(II)** oxide, CuO, form. On exposure to the air for several years, copper gradually becomes covered with a beautiful blue-green *patina*, which consists of a basic copper sulphate, $[Cu.3Cu(OH)_2]SO_4$. Nitric acid attacks copper, forming copper nitrate, $Cu(NO_3)_2$.

Below: The most important single use of copper is in the form of wire. Wire is made by a drawing operation on machines like these. It is drawn cold, for copper has excellent ductility even at ordinary temperatures.

Copper

COPPER ALLOYS

With zinc, copper forms a series of alloys called *brasses*. When brasses contain less than 36% zinc, they are malleable and ductile when cold, and they can be shaped cold with ease. Cartridge cases are often made with 70/30 brass (containing 30% zinc).

Brasses containing more than 36% zinc are much harder and stronger. Up to 36% zinc the alloy is a single-phase solid solution. Above 36% it is a two-phase solid solution, the phases being termed alpha and beta. One of the most widely used of these so-called alpha-beta brasses is 60/40 brass, better known as Muntz metal. Alpha-beta brasses are generally shaped hot.

Alloys of copper with tin are called *bronzes*. However, the term is also sometimes applied to copper alloys with other elements which contain no tin at all. The commonest bronze is that used for coinage. A typical composition would be 97% copper, 0·5% tin and 2·5% zinc. Alloys containing less copper and more zinc, known as *gunmetals*, are in widespread use—for marine applications for example. The so-called Admiralty gunmetal contains 88% copper, 10% tin and 2% zinc.

Phosphor bronzes usually contain less than 0·5% phosphorus with about 6% tin. Aluminium bronzes contain no tin at all, but up to about 12% aluminium. Silicon bronzes contain no tin either. All of these bronzes have excellent corrosion resistance and strength.

With nickel, copper forms an extensive range of alloys, the most familiar being the *cupronickels*, which contain up to 30% nickel. The 75% copper, 25% nickel alloy is used commonly for 'silver' coinage. So-called *nickel silver* containing copper, nickel and zinc, is widely used as a base metal in silver plating, the product being termed electroplated nickel silver. Among the other interesting copper alloys is *Monel*, which is a natural alloy obtained by smelting the nickel-copper Sudbury ore directly.

The first alloy in common use was bronze, a mixture of copper and tin. So common was it between about 3000 and 1000 BC that this period is referred to as the Bronze Age. Apart from being forged into tools and weapons, bronze was also worked into beautiful ornamental figurines such as this Chinese gilt bronze leopard.

Right: Some of the essential stages in the extraction and refining of nickel from Sudbury ore. It is a lengthy process because the ore is a mixture of other metals besides nickel. Valuable amounts of silver, gold and metals of the platinum group are recovered as by-products of nickel refining.

Nickel

Nickel is one of the metals of the first transition period of the Periodic Table, sandwiched between cobalt and copper. It has a combination of valuable properties. Attractive to look at, it takes a high polish and does not tarnish. It is highly ductile and malleable, so shaping it either hot or cold presents no difficulty. It is also magnetic.

One of the greatest uses of nickel is in electroplating. It is invariably used as an undercoat for chromium plating. Nickel is also a major alloying element in ordinary *stainless steel* (known as 18/8 because it con-

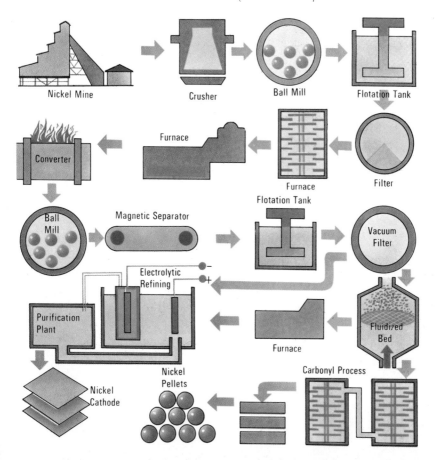

Nickel

sists of 18% chromium, 8% nickel, with the balance mainly iron). Another well-known use of nickel is in coinage. It is alloyed with copper to form the *cupronickel* alloys used today for so-called 'silver' coins. The cupronickel used for coinage contains about 25% nickel to 75% copper. Nickel and copper form many other useful alloys, which are characterized by high strength and corrosion resistance. *Monel metal*, which contains about 66% nickel, has many marine uses since it resists even sea-water corrosion.

One of the most interesting of its alloys is *invar*, which is a nickel-iron alloy containing 36% nickel. It has exceptionally low thermal expansion, so it is used for the balance wheels of watches, for measuring scales, and for precision instruments, whose performance would be affected by changing temperature.

Extraction and Refining

Nickel derives its name from 'Old Nick', a name for the devil that originated in Germany. Copper miners in the Hartz mountains in Saxony found ores that could not be reduced to workable copper and called them *Kupfernickel*, meaning devil's copper. It was later found that the contaminant was a new element—nickel. Nickel often occurs with copper in mineral deposits, together with iron. One of the main nickel minerals is *pentlandite*, a mixed sulphide with iron, $(Ni,Fe)_9S_8$. It occurs commonly with iron and copper sulphides, such as *chalcopyrite* $(CuFeS_2)$. The ore mined at Sudbury, Ontario, consists of a mixture of such minerals.

Extracting nickel from its ores is a lengthy process. Methods vary from plant to plant, depending on the exact composition of the ore, the availability of cheap electric power, the type of product desired (finely divided or slab), and so on.

An exquisite platinum goblet. Platinum is not widely used for making such large objets d'art because it is too expensive, It is, however, quite commonly used in jewellery.

Precious Metals

For thousands of years gold and silver have been prized, their scarcity enhancing their superlative qualities. Beautiful to look at, they are durable, slow to tarnish, and very easily worked. Gold in fact is the most ductile and malleable of all metals. It can be beaten into sheet, or *leaf*, so thin as to be transparent: 100 grams of solid gold will yield as much as 100 square metres of gold leaf.

Because of these qualities, silver and gold are *precious* metals. They are also called *noble* metals because they are not readily attacked by atmospheric gases and other common chemicals. They are often described, with copper, as *coinage* metals, because their outstanding durability once made them useful for making coins. But now no major countries

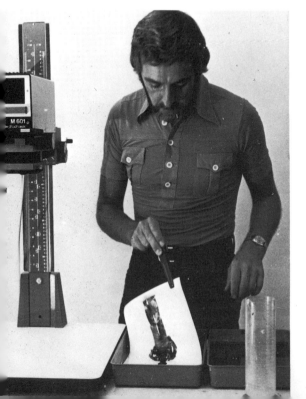

Left: The photographic industry is a major user of silver. Film is a mixture of silver halide grains suspended in gelatin on a cellulose-acetate base. The grains of silver halide — usually bromide with a little iodide — are sensitive to light. When light strikes the grains they undergo an invisible change which makes them chemically different from unexposed halide grains. When the film is treated with certain organic developing agents, the exposed grains are reduced to metallic silver. The unexposed grains are not affected. They are removed from the film by treatment with a fixing agent such as hypo — sodium thiosulphate. The developed image thus consists of tiny specks of silver.

Precious Metals

use silver or gold in their everyday coinage. The other major group of precious, noble metals contains platinum and the closely related palladium, rhodium, iridium, ruthenium, and osmium. This group is even scarcer than silver and gold.

Silver and Gold

Gold and silver are so chemically unreactive that they can be found native in the ground. Native gold is the most important source of the metal, though a great deal is also obtained stream bed, while the lighter fragments were washed away. Recovering the gold from placer deposits also involves the sorting action of flowing water. Hand panning, or swirling stream gravel and water around in a pan, was the earliest method, used for example by the miners of the gold-rush days in California (1849); the Yukon, Canada (1897); and Australia (1851). Today dredgers and vibrating sluice boxes are used to wash and sort placer deposits.

Gold is also mined underground. It is so precious that it is often worked at tremendous depths. In the Western Deep Levels Mine at

Right: A great deal of the gold produced by South Africa, the world's biggest producer, is mined deep underground. The gold-bearing ore body is broken up by explosives and by means of pneumatic rock drills.

Carltonville, South Africa, a depth of more than three kilometres has been reached. South Africa is the world's largest gold-producer, with an output of approximately 570,000 tonnes a year, about 70% of the world total. The gold-bearing ore is usually treated with sodium (or potassium) cyanide solution. The gold reacts with the solution and dissolves, as sodium (or potassium) dicyanoaurate, $Na[Au(CN)_2]$. When zinc is added to the solution, the gold is precipitated.

Both gold and silver conduct heat and electricity well, silver being superior in this respect to every other metal. But copper is generally used instead of silver for heat and electrical applications because it is much cheaper. Gold is used for long-lasting connections in electrical and electronic components. Dentists use it in alloys for tooth-filling. But its greatest use is in making

jewellery. Pure gold is too soft for this purpose, so it is alloyed. Commercial gold may also contain platinum, palladium, silver, copper, nickel, or zinc. The quality of the gold is measured in carats (or karats), a carat being one-twenty-fourth part: 18-carat gold, for example, contains 75% gold.

Similarly, for jewellery and tableware, silver is generally alloyed with copper to make it harder: so-called *sterling* silver contains at least 92·5% silver to 7·5% copper. Much silver tableware is not solid silver but silver plated over a base metal such as brass.

The Platinum Group

One of the finest catalysts for a wide range of chemical reactions is the precious metal platinum. Two processes in which this quality is utilized are the catalytic cracking of petroleum and the manufacture of nitric acid. Platinum is a soft, ductile and very dense metal (relative density 21·5), with exceptional resistance to corrosion; only aqua regia will attack it. Its melting point is high (1769°C). Apart from industrial uses, it is also used for jewellery.

Platinum is one of the commonest of a closely related group of metals, which also include palladium, rhodium, iridium, ruthenium and osmium. They all have a high melting point, excellent resistance to corrosion, and catalytic properties. They are transition elements which occur together in Group 8B of the Periodic Table, the group which also includes iron, cobalt and nickel. After platinum, palladium is the most useful metal of the group. It is cheaper than platinum; it is also lighter and less resistant to corrosion. Its main use is as a catalyst. But it is also used for long-lasting electrical contacts in telephone relays and similar communications devices.

Precious Metals

It has the interesting property of allowing hydrogen to diffuse through it at elevated temperatures.

Rhodium is a hard, exceptionally white metal, often used in electroplating. It forms a hard lustrous coating that does not tarnish. It is electroplated on electrical contacts where it guarantees long life. The other major use of the metal is as an alloying element in platinum for use in high-temperature thermocouples.

Iridium is very hard; has a high melting point ($2443°C$); and is the most corrosion-resistant of the group. It is also the heaviest of all elements (relative density 22·4) after osmium (relative density 22·5).

Platinum is widely used in the chemical industry for its catalytic properties and for its inertness and corrosion resistance. The machine below shows platinum gauze being woven on looms.

Tin, Lead and Zinc

These three common metals may not be in the front rank among metals, but they are nevertheless of great importance. Tin and lead are closely related chemically, being in Group 4A of the Periodic Table; zinc on the other hand is in Group 2B. All three metals have reasonable resistance to atmospheric attack and are often used for protective coatings.

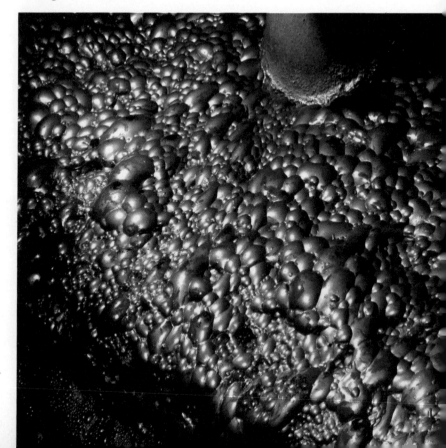

Crystals of the main lead ore, galena — lead sulphide. Galena, which is also called lead glance, has cubic crystals which have a dull, metallic lustre, resembling that of the parent metal. Galena is very heavy relative density 7·5). Most galena ore bodies also contain valuable traces of silver, and sometimes gold. These metals are valuable by-products of the refining operations.

The bubbles on a flotation tank in which zinc sulphide ore — zinc blende — is being separated from associated impurities. The mined ore is first ground fine and then treated with surface-active agents which preferentially wet the impurities present. The fine pure ore particles attach themselves to bubbles of air passing through the tank and collect in the froth at the top, which is floated off.

Perhaps their greatest importance lies in their use in alloys. Most notable are the alloys of tin and zinc with copper, they being known respectively as *bronze* and *brass* (see page 164). The bronze coins we use contain zinc as well as copper and tin. Lead and tin are the main ingredients of the long-used alloy *pewter*. They also form with antimony the alloy *type metal*. *Solders* used for joining metals are also alloys of tin and lead. Although tin and zinc are comparatively soft and weak metals, their alloys with copper are strong and hard. Lead too, is soft, and like the other metals can be shaped with ease. It is often included in alloys, such as *leaded bronze*, to improve machinability, acting effectively as a lubricant. Its density (relative density 11·3) is much greater than those of tin and zinc (7·3 and 7·1 respectively).

Extraction and Production

Tin occurs in nature primarily as the oxide, SnO_2, called *cassiterite* or *tinstone*. Cassiterite is found in lode deposits underground or in placer deposits in stream beds. It commonly occurs in placers because it is very heavy (relative density about 7)—almost as heavy as iron, in fact. Most of the world's tin comes from placer deposits in south-east Asia, especially Malaysia, which produces about a third of the annual world total of some 210,000 tonnes. Russia and Bolivia also are large tin producers.

The lead and zinc ores are sulphides, named *galena* or *lead glance* (PbS), and *zinc blende* or *sphalerite* (ZnS) respectively. The zinc ore is deceptively like the lead ore and, to add to the confusion, they are often found together in mineral deposits. Silver, and sometimes gold, are also often present in the ore body and are extracted as valuable by-products

Refractory Metals

Of all the metals, tungsten has the highest melting point (3380°C), a quality exploited in most of its applications. One of its most familiar applications is for the filaments of electric-light bulbs. Tungsten is one of several metals exploited for their temperature resistance. Others include molybdenum, which is closely related to tungsten; titanium and zirconium; niobium and tantalum. All these metals are transition metals in groups 4B, 5B or 6B of the Periodic Table. In general they have excellent corrosion resistance, but they tend to oxidize at high temperatures and are usually difficult to fabricate.

Tungsten used to be called *wolfram*, hence its chemical symbol, W. It is immensely strong and hard and is used to impart these properties, as well as temperature resistance, to alloys such as steel. Tungsten carbide, the combination of the metal with carbon, is one of the finest refractory materials. It is used for making high-speed drills, which lose little of their hardness even when very hot. The carbide is made by heating tungsten and carbon black at about 1500°C.

Molybdenum resembles tungsten in many respects, though its melting point is somewhat lower (2620°C). It is also softer and more ductile than tungsten, with which it forms a continuous range of solid solutions. A 50-50 tungsten-molybdenum alloy is sometimes used, which combines the higher refractoriness of tungsten with the greater ductility of molybdenum. Molybdenum is also widely used in alloys with steel and non-ferrous metals, where it confers hot strength and corrosion

Above: A titanium alloy vacuum chamber manufactured for the European Nuclear Research Organization (CERN). It is a complex assembly of 15 convoluted modules, made by a newly developed blow-forming process similar to that used for shaping glass. This technique is possible with this particular titanium alloy which has superplastic forming characteristics.

resistance. Molybdenum sulphide is one of its most widely used compounds, acting as a solid lubricant (rather like graphite). Molybdenum occurs naturally as the sulphide in the black mineral molybdenite.

Titanium (melting point 1680°C) is a metal increasingly used for construction, especially in the aircraft industry. It has the desirable properties of lightness combined with high strength and corrosion resistance. Its corrosion resistance depends, like that of aluminium, on the formation of a passive oxide film on the exposed surface.

Alkali Metals

One of the most reactive groups of elements in Nature are the alkali metals. They are so-called because they react with water to form hydroxides which are strong alkalis. The metals are, in order of increasing complexity, lithium (Li), sodium (Na), potassium (K), rubidium (Rb) and caesium (Cs). There is a sixth alkali metal, francium (Fr), but it is a very unstable radioactive element of which only a few grams ever exist in the Earth's crust at one time. The alkali metals form a very closely knit family, occupying Group 1A of the Periodic Table.

Sodium is by far the most important of the alkali metals, followed by lithium and potassium. Rubidium and caesium are relatively unimportant. They all, however, have much the same physical and chemical characteristics. For example, they are all lustrous metals, so soft that they can be cut with a knife.

The alkali elements themselves have few uses. Some lithium metal is used in lightweight alloys with magnesium. Sodium metal is used as a coolant in certain types of nuclear reactor; its high conductivity makes it a very effective heat-transfer medium. Sodium vapour is used to fill street lamps, which emit a characteristic yellowish-orange glow. Caesium also is used in vapour form in the atomic clock, the vibration of its atoms acting as a regulator.

Sodium and potassium compounds are by far the most important and most familiar of the alkali metal compounds.

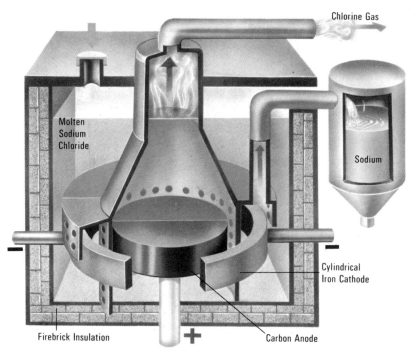

Above: The Downs cell, which is used for the commercial production of sodium metal. In the cell sodium forms at the cathode, while chlorine gas forms at the anode. The liquid sodium rises above the molten sodium chloride because it is less dense.

Sodium

One of the most abundant elements, sodium is, like all other alkali metals, too reactive to be found uncombined in Nature. It occurs widely in the Earth's crust, most notably as *salt* (sodium chloride, NaCl). Massive *rock-salt* deposits occur throughout the world; and the oceans contain nearly 3% sodium chloride.

Potassium

Potassium has a very similar chemistry to sodium, with compounds too numerous to be dealt with fully. Potassium is one of the essential elements plants need for proper growth, and for this reason the sulphate (K_2SO_4), chloride (KCl), and nitrate (KNO_3) of potassium are used as fertilizers.

Alkaline Earth Metals

This family of metals, in Group 2A of the Periodic Table, comprises beryllium, magnesium, calcium, barium, strontium and radium (which is radioactive). The most important of these metals by far are calcium and magnesium, whose compounds occur abundantly in the Earth's crust and are exploited on a large scale in industry.

The alkaline-earth metals are grey in colour and are moderately hard. They have relatively high melting points, ranging from 1280°C for beryllium to 850°C for calcium. Like all metals they conduct heat and electricity well. Industrially, the most important of the metals is magnesium. It is a very lightweight metal (relative density 2·7). It is alloyed with aluminium, zinc, and manganese to make lightweight alloys for use in aircraft. These alloys are readily shaped by normal methods such as forging, casting and machining. Indeed magnesium is the easiest of all metals to machine.

Magnesium

In Nature magnesium is never found uncombined, for, like all the alkaline earths, it is too reactive. Among the minerals containing magnesium are *magnesite*, $MgCO_3$; *dolomite*, $CaCO_3.MgCO_3$; *kieserite*, $MgSO_4.H_2O$; *carnallite*, $MgCl_2.KCl.6H_2O$; and many silicate minerals, including *talc* and *asbestos*. Seawater contains an abundance of magnesium in the form of soluble chloride, and extraction plants now process seawater to obtain magnesium metal.

Below: A magnesium alloy casting used on the RB-211 turbofan engine which powers the Lockheed Tristar airbus.

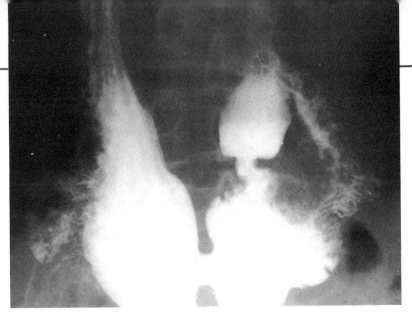

Above: An X-ray photograph of the human digestive system. Normally this would not be visible on an X-ray. It shows up here because the patient has had a 'barium meal' — a liquid containing barium sulphate. This mineral is highly opaque to X-rays.

Calcium

The fifth most abundant of all the elements, calcium comprises nearly 4% of the Earth's crust. It is most commonly found as the carbonate, $CaCO_3$, in massive sedimentary rocks such as *limestone* and *chalk*. These deposits were formed hundreds of millions of years ago from the remains of marine organisms, such as coral and shellfish, which extracted calcium from the ancient seas to build up their shells.

Strontium and Barium

Although not very important, these alkaline earths do have one or two useful compounds. Strontium chlorate, $Sr(ClO_3)_2$, and nitrate, $Sr(NO_3)_2$, are incorporated in fireworks, flares, and other pyrotechnic devices, where they impart a vivid crimson colour.

Barium sulphate, $BaSO_4$, which occurs naturally as *barytes*, or *heavy spar*, is the white substance given as a barium meal to patients in hospital having stomach and intestinal X-rays taken. The mineral is particularly opaque to X-rays, and causes the inner organs to show up on the X-ray film.

The Atmosphere

The Earth is enveloped by a layer of gases which we call the atmosphere. This layer is effectively only a few hundred kilometres thick. If the Earth were an apple then the atmosphere would be no thicker than the peel. Yet the atmosphere is of the most profound importance. By providing oxygen for respiration it allows an abundance of life to exist on Earth. It absorbs harmful radiation from the Sun and acts as a blanket to equalize day-time and night-time temperatures by trapping the Sun's day-time heat.

Oxygen makes up only about 21 per cent by volume of the atmosphere. Most of the remainder is nitrogen (78%). There is also argon (0·93%), carbon dioxide (0·03%), neon (0·002%), and traces of many other gases, including sulphur dioxide, carbon monoxide and nitrogen oxides. The latter are more prevalent in industrial areas and are the main sources of atmospheric pollution. Water vapour, too, is always present in the atmosphere in varying quantities.

The Pressure of the Air

The Earth's gravity not only prevents the gases of the atmosphere from escaping into space; it also compresses them against the surface of the planet. In fact, half of the atmosphere's total mass lies within 5·5 kilometres of sea level. The pressure of the atmosphere at sea level is about one kilogram per square centimetre, termed *one atmosphere*. The pressure drops markedly with height above sea level. At 10,000 metres (little more than the height of Everest) the pressure is less

Right: The envelope of the Earth's atmosphere is several hundred miles thick yet 90 per cent of air is concentrated in the lowest 10 miles (16 km). Though the upper layers cannot support life, they protect the planet from harmful solar radiation and incoming meteoroids.

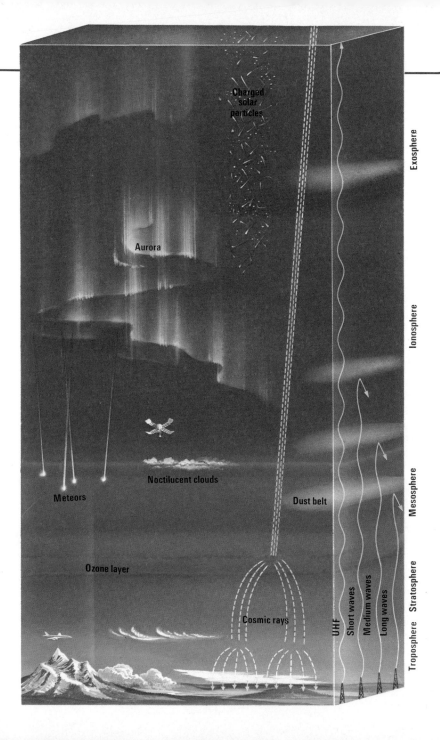

The Atmosphere

than one-third of an atmosphere. Thereafter, the pressure falls even more rapidly until at a height of about 800 kilometres there are only a few gas molecules remaining, mainly of hydrogen and helium. The atmosphere gradually merges with space.

Layers of Air

The atmosphere can be divided into five main regions: the *troposphere* (the lowest layer), the *stratosphere*, the *mesosphere*, the *ionosphere* and the *exosphere*.

The troposphere is the part of the atmosphere in which we live and to which weather is confined. Temperature decreases steadily with height in this region, falling to a freezing $-55°C$ at the *tropopause*, a thin boundary layer marking the top of the troposphere. The height of the tropopause varies from 16 kilometres over the equator to 8 kilometres over the polar regions. Icy, high-speed winds called *jet streams* blow in thin ribbons around the world at this height. Their speed can exceed 240 kilometres per hour.

Except on short flights, airliners climb through the tropopause into the calmer air of the stratosphere. Wind speeds decrease with height in the stratosphere, but the temperature *increases*. In the short space of 25 kilometres it rises to around freezing point. This warming effect is due to the action of ultra-violet rays from the Sun. As ultra-violet radiation changes oxygen into ozone, heat is produced.

The Upper Atmosphere

Above the ozone layer lies the mesosphere in which the temperature falls to $-70°C$ at its boundary with the ionosphere (about 80 kilometres above the Earth).

The ionosphere is so called because the

Above: A Nimbus weather satellite picture of Europe. It shows the skies free of cloud practically everywhere. Only over Ireland has cloud built up.

The Atmosphere

atoms of gas there are *ionized* (turned into electrically charged atoms, or *ions*) by the passage of ultra-violet rays, X-rays, and streams of atomic particles from space. Heat is produced in the process and the temperature in the ionosphere consequently rises with height to more than 20,000°C. The ionosphere is vital to life on Earth. It is at this level that most of the harmful radiation from space is absorbed. The upper levels of the ionosphere also cut down the amount of infra-red ('heat') radiation reaching the Earth's surface. The ionosphere is also important in radio communications because it contains layers which are able to reflect certain radio waves. Before the advent of communications satellites it was these reflective layers alone that enabled radio transmissions to be picked up around the world.

The upper ionosphere contains a high proportion of nitrogen, but very little oxygen. Instead, a great deal of ozone is found. Nearer to the Earth's surface, the E-region contains more nitrogen and more oxygen, but the proportion of ozone is correspondingly less.

The actual chemical content of the ionosphere changes from day to day, varying with the radiation bombardment from space. The number of free ions and electrons in the ionosphere changes with the time of day as well. At night, when the Sun's rays are not falling on the ionosphere, the charged particles tend to recombine in the lower parts of the layer. But at higher levels the atmosphere remains substantially ionized because the lower density of ions decreases the chance of collisions with free electrons and with each other.

Above 500 kilometres lies the exosphere, the tenuous outermost layer of the atmosphere which merges imperceptibly with space.

Oxygen

Oxygen is the commonest element on Earth. Nearly all rocks and clays in the Earth's crust contain a high proportion of it in chemical combination, as do many minerals. The oceans are composed of about 86 per cent by weight of oxygen. It is also an essential constituent of all living things; more than 70 per cent by weight of the human body is oxygen in chemical combination.

Oxygen is a colourless, odourless gas that makes up about 21 per cent of the atmosphere. The proportion remains fairly constant because oxygen consumed during respiration and combustion is continually replaced by the photosynthetic processes of plants. Photosynthesis is the process by which plants form organic compounds from carbon dioxide (CO_2) and water (H_2O) using energy in the Sun's rays. Respiration is the chemical process by which animals use oxygen to break down food into energy. Photosynthesis and respiration are coupled by the constant flow through the biosphere of the raw materials for their reactions. Respiration uses up oxygen and produces carbon dioxide; photosynthesis uses up carbon dioxide and produces oxygen.

The Manufacture of Oxygen

Oxygen is manufactured by separation from liquid air in a process known as *fractional distillation*. Air is compressed to 10 atmospheres, cooled, and filtered to free it from dust, carbon dioxide, and water. Then it is cooled again. Further compression to 200 atmospheres and further cooling liquefies the air. Nitrogen,

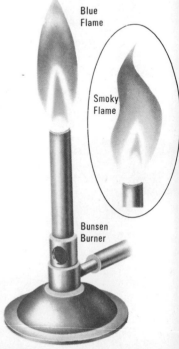

Below: The characteristic hot blue flame of a Bunsen burner with the air hole wide open. Inset is the smoky flame with the air hole closed.

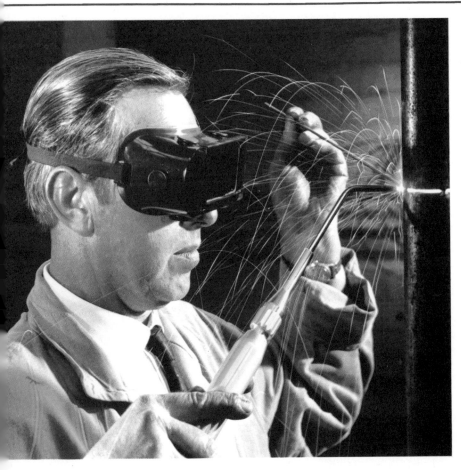

Above: Oxyacetylene welding. Oxygen and acetylene are fed to the welding torch to produce a flame burning at about 3500°C.

which is more volatile, is allowed to boil away (at a temperature of −196°C), leaving a liquid containing about 40 per cent oxygen, which boils at −183°C. This is transferred to a further distillation process which produces liquid oxygen that is about 98 per cent pure. The liquid oxygen is drawn off and stored in cylinders or other storage vessels under pressure. Liquid oxygen is pale blue in colour and strongly magnetic. Countries with an

Oxygen

abundance of hydro-electric power manufacture oxygen by the electrolysis of a dilute solution of sodium hydroxide (NaOH).

Oxygen is an extremely reactive element; it combines directly with many elements and compounds, particularly when they are heated. These reactions are often highly *exothermic* (giving out heat) and are of great industrial and domestic importance as sources of energy: for example, the burning of fuels such as coal, coke, oil, and gas. Industrially, oxygen is used in oxy-acetylene and oxy-hydrogen welding. Steel plants also consume such enormous amounts of oxygen that they often have their own oxygen production plants. Liquid oxygen is used as a major propellant in rockets and missiles.

Combustion

Nearly all oxidations are exothermic, especially those where oxygen combines directly with other elements or compounds. Even rusting, the oxidation of grey coloured iron or steel to the familiar red oxide, gives out some heat, but it is immediately dissipated to the surroundings. More obvious examples of exothermic oxidations are those involved in burning, or combustion.

Combustion is defined as a chemical reaction or series of chemical reactions in which a substance combines with oxygen producing heat, light and flame.

The Bunsen burner used in laboratories for heating materials demonstrates how flames differ with the amount of oxygen consumed in the reaction. With the air valve fully closed, the gas burns with a yellow or luminous flame. This is because oxygen can only mix with the gas from the outside of the flame, leaving unburnt particles of carbon glowing in the flame cone. When the air valve is opened,

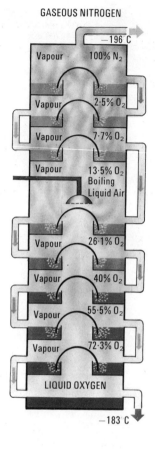

Oxygen

A diagram of a fractional distillation tower where the lower-boiling nitrogen is separated from the oxygen in liquid air. Boiling liquid air is introduced into the column half-way up. Vapour travels up through the column and passes through the liquid at each level, which is maintained at a certain temperature. As vapour passes up through the column it becomes richer in nitrogen. As the liquid spills over going down the column, it becomes richer in oxygen.

oxygen mixes thoroughly with the gas and burns it more efficiently. This produces the familiar blue flame.

Combustion is used to drive machines, such as internal combustion engines. The car engine uses a mixture of petrol and air as a fuel. As the piston rises, compressing the mixture, the spark from the plug causes oxygen molecules to dissociate into active oxygen particles. These particles immediately react with the long carbon-chain compounds in the petrol, breaking them down rapidly with the evolution of energy. The energy causes the breakdown of more oxygen molecules, and therefore more fuel, as the flame front spreads across the combustion chamber. The principle is the same as that used in lighting the domestic fire or firing an industrial boiler.

CARBON OXIDES

Carbon monoxide (CO) is a colourless, odourless gas which is very poisonous, even in minute concentrations. It is present, for example, in the exhaust fumes from automobile engines, resulting from the incomplete combustion of hydrocarbon fuel. In industry carbon monoxide is an important chemical. It is usually made by passing air or steam through a bed of white-hot coke. When steam is used, hydrogen is also formed. The gas so produced, often called blue gas or water gas, is an important heating gas. Carbon monoxide is also a major ingredient of coal gas. It burns in air with a blue flame, forming carbon dioxide.

Carbon monoxide can be made to combine with hydrogen to form methyl alcohol, or methanol, and this is a very important industrial process. Carbon monoxide is a strong reducing agent. It also combines directly with some metals to form volatile carbonyls. This property is utilized in the Mond process for nickel refining.

Carbon dioxide (CO_2) is present in the air in varying amounts, averaging about 0·03 per cent by volume. It is formed when any carbon-containing material burns completely. It is also produced during fermentation processes, and is given out by animals when they breathe. Plants use carbon dioxide to make food during photosynthesis. Large volumes of carbon dioxide are produced in various industrial processes, including the roasting of limestone to make quicklime (calcium oxide). In the laboratory carbon dioxide can readily be obtained by the action of dilute acids on limestone. Carbon dioxide can easily be liquefied under pressure, and if the pressure is released, a fluffy, snow-like solid forms, called dry ice. It is a convenient portable cooling agent which leaves no mess because it sublimes without melting.

The Rare Gases

As the name implies, the rare gases occur naturally in only very small quantities. The six gases, helium, neon, argon, krypton, xenon and radon are marked by their lack of reactivity. It was this property that led to them being called the 'inert gases' when they were first discovered. In fact, they are not totally inert and in recent years scientists have succeeded in making some compounds.

The reason for the low reactivity of the rare gases lies in the structures of their atoms. The electron shells are all filled. This means that they do not need to combine with other atoms to try to fill the shells. Most gases have two atoms in their molecules, but the rare gases have only one. In other words, they are *monatomic*. The 'fully filled' electronic structure is sometimes known as the noble gas structure, and the rare gases are also known as the *noble gases*.

The rare gases occupy about one per cent by volume of the air, argon being by far the most plentiful. Some natural spring waters contain small amounts of dissolved helium, neon and argon, and natural gas, particularly that found in the United States, may contain up to two per cent of helium. Helium is also produced in small quantities during the radioactive decay of uranium. It can be obtained from minerals containing uranium by strong heating. Astronomy has revealed the presence of vast amounts of helium in stars, the older stars containing the most. It is thought that the thermonuclear fusion of hydrogen into helium is the principal source of energy in stars.

Below: Liquid helium being used in low-temperature, or cryogenic, research. Its boiling point (-268.9 °C) is the lowest of any element, and is only 4.2 °C above absolute zero.

Above: Argon is used as an inert gas in electric-light bulbs, In this picture argon is being injected in the fourth bulb from the right. The second bulb from the right is being sealed.

Preparation and Properties

The main source of helium is natural gas, which consists predominantly of nitrogen and hydrocarbons. The natural gas is liquefied by cooling under pressure and the residual helium, which liquefies at a lower temperature, is drawn off and passed over cooled, activated charcoal, which removes any traces of heavier gases. Helium has several interesting chemical and physical properties. It is very light, being second only to hydrogen, and for this reason can be used for filling balloons and airships. It is also much safer for these purposes than hydrogen because it does not burn.

Nitrogen

Nitrogen is a colourless, odourless gas that makes up four fifths of the air we breathe. It does not burn and hardly dissolves in water at all. Nitrogen molecules in the air contain two atoms. These diatomic molecules are extraordinarily stable.

Nitrogen in Nature

Nitrogen is found in all organisms and is a vital part of proteins and other chemicals in the body. Air-breathing animals cannot use nitrogen directly from the air. They obtain it in their food. Some plants, however, have the ability to 'fix' nitrogen, combining it with other chemicals in complicated chemical reactions. Other plants absorb it in chemical form directly from the ground. When animals and plants die and rot, their nitrogen content either returns to the ground in chemical form or is given off into the atmosphere, usually as ammonia (NH_3).

Below: The natural nitrogen cycle. Nitrogen is essential to plant and animal life, being present in the protein of body tissues. Plants acquire their nitrogen through the absorption of nitrates from the soil. The nitrates are replenished naturally as a result of electrical discharges in the atmosphere, which result in the production of nitric acid. The rain washes the nitric acid into the soil, where it forms nitrates. Nitrates are also formed by the oxidation of ammonia produced by the rotting bodies of animals and plants. Bacteria in the soil bring about the oxidation. Denitrifying bacteria convert some nitrates into free nitrogen.

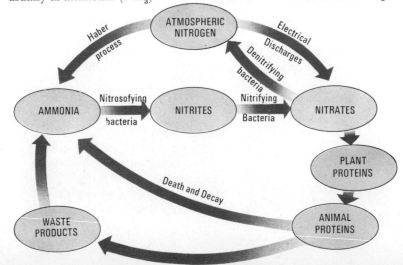

As world demand for food has risen, it has been found that there is not enough readily-available nitrogen in the soil, so the natural levels of the element are supplemented by the use of nitrogen-containing fertilizers.

Nitrogen in Industry

Nitrogen is also found in naturally occurring chemicals such as Chile saltpetre, a nitrate which was at one time the main industrial source of nitrogen gas. Now nitrogen is extracted from the air. Air is a mixture of

Ammonia

Most of the nitrogen produced industrially is used in the manufacture of ammonia. Ammonia used to be made from coal, but in the early years of this century it was found that if nitrogen and hydrogen are heated together at high pressure, they combine to give ammonia. This process, known as the *Haber process*, has largely superseded other methods of manufacturing ammonia. High temperatures (500°C) and pressures (1000 atmospheres) are needed in this process because of the stability, or inertness, of the nitrogen molecule.

A diagram representing the Haber process for the manufacture of ammonia from nitrogen and hydrogen. Three volumes of hydrogen and two volumes of nitrogen will combine to form two volumes of ammonia, the reaction being reversible. This reaction takes place at a reasonable rate at a temperature of about 500 °C, under a pressure of about 1000 atmospheres, and in the presence of an iron catalyst. Only part of the reacting gases combine as they pass through the catalyst chamber. The gas leaving the chamber is cooled, whereupon ammonia condenses. The uncombined nitrogen and ammonia are recirculated into the catalyst chamber.

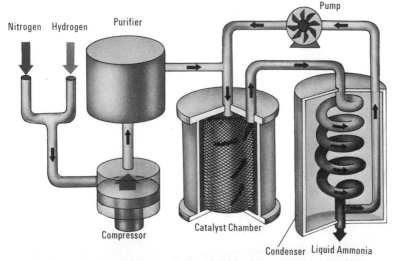

Water

Water is the most important compound on Earth. It can exist in three states—vapour, liquid and solid (ice). About $2\frac{1}{2}$ per cent of the Earth's water is frozen in glaciers and ice sheets. Most of the rest is liquid. It covers about 70 per cent of the Earth's surface, most of it being water in the oceans. Under the heat of the Sun, water is evaporated from the oceans. When the humidity of the air increases beyond saturation point (the maximum amount of water vapour the air can contain at a particular temperature), droplets of water form. These mass together as clouds. When the droplets become too large to be held in the air they fall as rain. Most of the

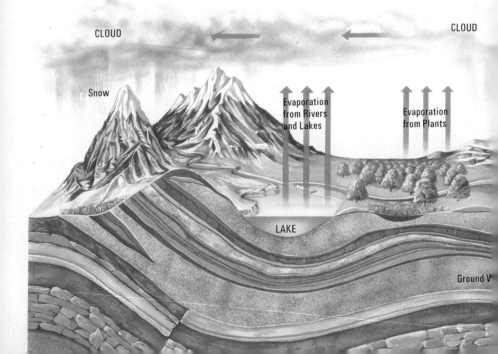

rain that falls on land runs into streams, rivers and lakes, and finally back into the sea.

Some water soaks into the soil and becomes *ground water*. Much of this is taken up by plants. Plants, like all other forms of life, depend upon water. It has been estimated that about 450 kilograms of water are needed to produce 500 grams of plant material. Animals also consume large amounts of water.

The Structure of Water

Water is made up of hydrogen and oxygen. Its formula—H_2O—has been established by many experiments. In the laboratory this can be done by the electrolysis of water to which has been added a little sulphuric acid. Pure water does not conduct electricity so the acid is added to complete the electrical circuit. The current breaks up water, giving two volumes of hydrogen for every one of oxygen.

In water the negative end of one molecule tends to attract the positive hydrogen nucleus of a neighbouring molecule, forming what is called a *hydrogen bond*. Hydrogen bonding explains the frequently anomalous behaviour of water compared with chemically similar compounds, such as hydrogen sulphide, H_2S. Hydrogen bonding causes the water molecules to stick together, giving water relatively high viscosity and high melting and boiling points (0°C and 100°C). By rights, compared with similar compounds, water should be a gas at room temperature. In the liquid state molecules of water are packed relatively close together, being at their closest (water being most dense) at about 4°C. In the solid state the asymmetric water molecules organize themselves into an ordered crystal lattice and occupy more space. Solid water (ice) is less dense than liquid water, or to put it another way, water expands on freezing.

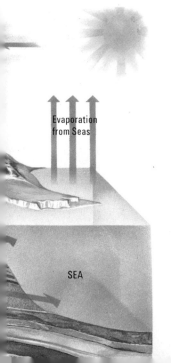

Below: There is a continual exchange of water between the Earth and the atmosphere, this forming what is called the water cycle. In this cycle water evaporates from the seas, rivers and lakes, and from growing plants. The vapour condenses into clouds in cool air, and returns to the ground as rain, snow or some other form of precipitation.

Hydrogen

Hydrogen is the lightest of all the known elements. By itself it forms a colourless, odourless gas. But very little free hydrogen is found on Earth. Small amounts are found in natural gas. The atmosphere contains less than one part per million of free hydrogen, and most of this occurs in the upper atmosphere. Large amounts of hydrogen exist in the sun. There, the nuclear fusion of hydrogen atoms into other elements provides vast quantities of energy.

The discovery of the structure of the hydrogen atom has made a unique contribution to our understanding of the structures of other elements. Normally, the hydrogen atom contains one electron and a nucleus consisting of one proton, but tiny quantities of other isotopes have been found. Hydrogen molecules are *diatomic* (having two atoms), and the element reacts with other substances very easily. Liquid hydrogen boils at $-253°C$.

Hydrogen occurs naturally combined with oxygen, carbon, and other elements in water, natural gas, and other organic substances. Water is the most important oxide of hydrogen, being essential for all plant and animal life. But hydrogen can form another oxide called hydrogen peroxide (H_2O_2), which is widely used as a bleaching agent.

The Manufacture of Hydrogen
Large amounts of hydrogen gas are used industrially. In countries with large supplies of natural gas, hydrogen is extracted from methane (CH_4).

Hydrogen is undoubtedly the most abundant element in the universe, being present in the stars and the gas clouds, or nebulae, between the stars. This picture shows the Trifid nebula in the constellation of Sagittarius. About 12 light-years across, it lies about 3000 light-years away.

Right: A battery of reactor vessels used in the hydrogenation of animal and vegetable oils into margarine. Hydrogenation converts the liquid oils into solids.

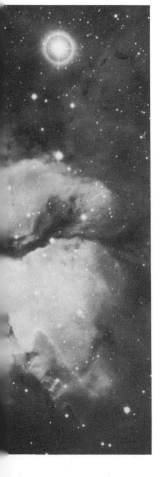

Great quantities of hydrogen are used in the Haber process for the manufacture of ammonia, and some, mixed with oxygen, is used in welding. Hydrogen was once used in airships and balloons because it is lighter than air. But after a series of airship disasters caused by the hydrogen catching fire, helium, which is non-flammable, was substituted.

PERIODIC TABLE OF THE ELEM

A modern version of the Periodic Table of the elements, giving their symbols, atomic numbers and atomic weights. The atomic weight is the mass of the atom on a scale that assigns the isotope carbon-12 an atomic weight of 12. Most of the atomic weights given are not whole numbers because an element is usually made up of two or more isotopes, each with a different mass. Iron, for example, is naturally made up of four main isotopes having masses of 54, 56, 57 and 58. The atomic weight given for iron reflects the relative abundance of its isotopes in Nature.

PERIODS

	1A	2A	3B	4B	5B	6B	7B	
1	1.01 **H** 1							
2	6.94 **Li** 3	9.01 **Be** 4						
3	22.99 **Na** 11	24.31 **Mg** 12						
4	39.09 **K** 19	40.08 **Ca** 20	44.96 **Sc** 21	47.9 **Ti** 22	50.94 **V** 23	52 **Cr** 24	54.94 **Mn** 25	55.84 **Fe** 26
5	85.47 **Rb** 37	87.62 **Sr** 38	88.91 **Y** 39	91.22 **Zr** 40	92.91 **Nb** 41	95.9 **Mo** 42	97 **Tc** 43	101 **Ru** 44
6	132.91 **Cs** 55	137.34 **Ba** 56	Rare Earths Lanthanides 57-71	178.4 **Hf** 72	180.95 **Ta** 73	183.8 **W** 74	186.2 **Re** 75	190.2 **Os** 76
7	223 **Fr** 87	226.03 **Ra** 88	Actinides 89-103	104	105	106		
GROUPS	1A	2A	3B	4B	5B	6B	7B	

KEY
- Light Metals
- Heavy Metals
- Non-metals / Submetals
- Rare Gases

138.91 **La** 57	140.12 **Ce** 58	140.91 **Pr** 59	144.2 **Nd** 60	(147) **Pm** 61

227 **Ac** 89	232.04 **Th** 90	231.04 **Pa** 91	238.03 **U** 92	237.05 **Np** 93

							7A	0
							1.01	4
	Atomic Weight						H	He
			3A	4A	5A	6A	1	2
			10.81	12.01	14.01	16	19	20.17
	Atomic Number		B	C	N	O	F	Ne
			5	6	7	8	9	10
Chemical Symbol			26.98	28.08	30.97	32.06	35.45	39.94
			Al	Si	P	S	Cl	Ar
	1B	2B	13	14	15	16	17	18
58.7	63.54	65.38	69.72	72.5	74.92	78.9	79.9	83.8
Ni	Cu	Zn	Ga	Ge	As	Se	Br	Kr
28	29	30	31	32	33	34	35	36
106.4	107.87	112.4	114.82	118.6	121.7	127.6	126.9	131.3
Pd	Ag	Cd	In	Sn	Sb	Te	I	Xe
46	47	48	49	50	51	52	53	54
195	196.97	200.5	204.3	207.2	208.98	209	210	222
Pt	Au	Hg	Tl	Pb	Bi	Po	At	Rn
78	79	80	81	82	83	84	85	86
	1B	2B	3A	4A	5A	6A	7A	0

SERIES — THE RARE EARTHS

96	157.2	158.93	162.5	164.93	167.2	168.93	173	174.97
	Gd	Tb	Dy	Ho	Er	Tm	Yb	Lu
	64	65	66	67	68	69	70	71

SERIES (93-103 are man made)

	247	247	251	254	257	257	255	256
	Cm	Bk	Cf	Es	Fm	Md	No	Lr
	96	97	98	99	100	101	102	103

A–Z of Science

absolute temperature Temperature related to *absolute zero*. See *kelvin*.

absolute zero Lowest temperature possible in theory; zero on absolute scale is −273·15°C. The lowest temperatures that can be reached in practice are within a few millionths of a degree above absolute zero. See also *kelvin*.

absorption Penetration of a substance into the body of another; e.g. a gas dissolving in a liquid. See also *adsorption*.

acceleration Rate of change of velocity; measured in distance per second per second.

accumulator Device for storing electricity, in which an electric current is passed between two plates in a liquid such as sulphuric acid.

acid Chemical substance that when dissolved in water produces hydrogen ions, which may be replaced by metals to form salts.

adhesion Attractive force between molecules of a substance that makes them stick together.

adsorption Phenomenon by which a gas or liquid becomes concentrated on the surface of a solid. See also *absorption*.

alkali *Base* consisting of a soluble metal hydroxide. Alkali metals, such as sodium and potassium, form *caustic alkalis*.

alloy Metal composed of more than one element; e.g. dentist's amalgam is an alloy of mercury (70%) and copper (30%).

alpha-rays (α-rays) Streams of helium nuclei emitted by some radioactive elements.

alternating current Electric current that rapidly decreases from maximum in one direction, through zero, and then increases to maximum in the other direction.

ammeter Electrical instrument for measuring current; a pointer moves over a scale graduated in amperes.

ampere (A) Unit of electric current equivalent to flow of 6×10^{18} electrons per sec (i.e. 6 million million million electrons). It is the basic SI unit of current; named after André Marie Ampère.

amplifier An electronic device that increases the strength of a signal (e.g. radio waves) fed into it by using power from another source.

amplitude The maximum value of a regularly varying quantity (something going back and forth) during a cycle; e.g. the maximum value of an alternating current.

angstrom unit (Å) Unit of length formerly used for the measurement of wavelengths of light; equivalent to 10^{-10} metre.

anode Positive electrode through which electric current enters an electrolytic cell or a vacuum tube.

Archimedes' principle When a body is immersed or partly immersed in a fluid, the apparent loss in weight is equal to the weight of the fluid displaced.

armature Rotating part of electric motor or dynamo, consisting of coils of wire.

atom Smallest fragment of an element that can take part in a chemical reaction. Consists of central *nucleus* (made up of *protons* and *neutrons*) surrounded by orbiting *electrons*. Number of protons (equal to number of electrons) is called the *atomic number*. Total mass of all the atomic particles is called the *relative atomic mass*. See also *isotope*.

atom smasher General name for any machine, such as a *cyclotron*, that accelerates atomic particles to sufficiently high speeds to split atoms.

atomic number See *atom*.

Right: A chain of thorium atoms photographed with a special high-power microscope. Each white dot is one atom. Thorium is a heavy, radioactive metal.

atomic weight Old name for relative atomic mass.

attraction Any force that tends to pull objects closer together. Forces of attraction include electrostatic, gravitational, magnetic, and inter-molecular forces.

base Substance that reacts chemically with an *acid* to form a *salt* and water.

battery Device that converts chemical energy into electrical energy; most batteries consist of a series of electric cells and provide a source of *direct current*.

beta-particles Stream of electrons emitted by some radioactive elements.

boiling point Temperature at which liquid turns into vapour throughout its bulk; the vapour pressure of the liquid then equals the external pressure on it.

bond Chemical link between two atoms in a *molecule*.

Boyle's law At constant temperature, the volume of a gas is inversely proportional to its pressure; i.e. the higher the pressure, the smaller the volume.

calorie Unit of heat equal to the amount needed to raise the temperature of 1 gram of water through 1 degree C; 1 calorie = 4·2 joules; the *kilocalorie* (=1000 calories), written *Calorie* (capital C), is used for food values.

candlepower Luminous intensity of a light source in a given direction, expressed in *candelas*.

candela The SI unit of luminous intensity.

capacitance Property of a capacitor that enables it to store electric charge; the numerical value of capacitance is measured in *farads*.

capacitor Circuit element, consisting of an arrangement of conductors and insulators that can store electric charge when voltage is applied across it; formerly called a *condenser*.

capillarity Phenomenon that makes liquid rise up a narrow space, as in a fine-bore tube or between two sheets of glass. See also *surface tension*.

catalyst Substance that markedly alters the speed of a chemical reaction, without appearing to take part in it.

cathode Negative *electrode* through which electric current leaves an electrolytic cell or a vacuum tube.

cathode rays Electrons flowing from the *cathode* of a discharge tube or valve.

Celsius Temperature scale on which 0°C is the melting point of ice and 100°C is boiling point of water; same as centigrade. To convert Celsius to Fahrenheit, multiply by 9, divide by 5, and add 32. See *Fahrenheit*.

centigrade See *Celsius*.

centrifugal force Fictitious force that appears to act on an object moving in a circular path.

centrifuge Machine that rotates at high speeds to separate solids from liquids or to separate liquids of different densities.

centripetal force Force that acts inwards on an object moving in a circular path.

cgs system Metric system of units based on centimetre, gram, and second; now largely replaced by SI units.

Charles's law At constant pressure, the volume of a gas is proportional to its absolute temperature; i.e. the higher the temperature, the greater the volume.

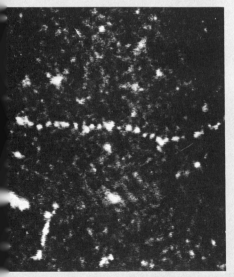

201

A-Z of Science

chemical equivalent Mass of an element that will combine with or replace 1 gram of hydrogen or 8 grams of oxygen. Of an acid, the mass containing 1 gram of replaceable hydrogen; now obsolete.
chloride Salt of hydrochloric acid.
circuit, electrical The complete path taken by an electric current.
cohesion Force of attraction between molecules that holds a liquid or a solid together. This cohesive force decreases with the rise in temperature.
colloid Substance in solution in a *colloidal state* – a system of particles in a medium with different properties from the true solution because of the larger size of particles; e.g. a solution of starch.
combustion (burning) Chemical reaction in which a substance combines with oxygen and gives off heat and light, and burns with a flame.
compound Substance consisting of two or more *elements* in chemical combination in definite proportions; e.g. water (hydrogen and oxygen) and salt (sodium and chlorine).
concave Curving inwards. A concave lens is thinner at the middle than at the edges.
concentration Amount of a substance expressed as mass in a given volume; concentrations of chemical solutions are generally expressed in moles per cubic metre.
condensation Change of vapour into liquid that takes place when pressure is applied to it or the temperature is lowered.
condenser, electrical See *capacitor*.
conduction, thermal The process of transmitting heat from molecule to molecule; e.g. a metal rod held in a flame at one end will transmit the heat to the other end.
conductor Substance that permits the flow of electricity; e.g. metals, such as copper.
convection Transfer of heat by means of the movement of heated matter from one place to another. It takes place in liquids and gases through the actual movement of the fluid.

convex Curving outwards. A convex lens is thinner at the edges than at the middle.
corrosion Slow chemical breakdown, often by oxidation of metals, by the action of water, air, or chemicals such as acids.
cosmic rays High-energy radiation, mainly in the form of charged particles striking the Earth from outer space.
coulomb (C) Unit of quantity of electricity, defined as the quantity transferred by 1 ampere in 1 second. The coulomb is the SI unit of electrical charge; named after Charles Augustin Coulomb.
critical temperature Temperature above which a gas cannot be liquefied no matter how great the pressure.
crystal Substance that has been solidified in a definite geometrical form. Some solids do not form crystals, and are said to be *amorphous*. The geometrical patterns are called *lattices*. Crystals are classified according to the structure of their lattices and the bonds holding them together. See also *crystallization*.
crystallization Process in which a regular solid substance (crystal) forms from molten mass or solution. See also *hydrate*.
curie Measure of *radioactivity* defined as the amount of a radioisotope that decays at a rate of 3.7×10^{10} disintegrations per second. Named after Marie Curie.
current, electric Flow of electrons along a conductor; measured in *amperes*.
cyclotron Machine for accelerating atomic particles to high speeds; particles follow spiral path in a magnetic field between two D-shaped electrodes. The particles build up energies of several million *electron-volts*.
decay Natural breakdown of a radioactive element. See *radioactivity*.
decibel Unit for comparing power levels or sound intensities; tenth of a *bel*.
density Mass per volume.
density, relative The density of a substance compared with that of water; defined as the ratio of the density of a substance to that of water at a temperature of

A-Z of Science

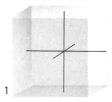

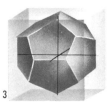

Minerals within the same crystal system may look unalike but still possess the same kind of symmetry. The cubic system has three equal axes at right angles (1), and many crystals in this system are cubic in shape. But by cutting corners from the cube, other shapes occur. An octahedron is first obtained, as in gold crystals (2), and then more complex shapes, as in pyrite (3) and leucite (4).

CRYSTAL SYSTEMS

Although crystals may appear to come in all kinds of shapes, this is because they are often composed of clumps of smaller crystals. In fact, there are only a few basic shapes. These shapes are classified on the basis of their axes of symmetry, for the rows of atoms or molecules in a crystal, and therefore the faces, always align themselves with an axis of symmetry.

Hexagonal System. There are four axes, three of which are equal in length and meet at 60°. The fourth axis is at right angles and is of unequal length. The crystals are often six-sided. They include apatite (right above), zincite (right below) and red cinnabar and white quartz (below).

A-Z of Science

4 °C. It is equal to the density expressed in grams per cubic centimetre. Formerly known as *specific gravity*.

desiccation The removal of moisture from a substance (drying).

dielectric Material that acts as an electrical insulator; i.e. it does not conduct electricity. In dielectrics, the electrons are bound so tightly to their atoms that few of them can move; e.g. rubber, glass, certain plastics.

diffraction The spreading out of light by passing it through a narrow slit or past the edge of an obstacle.

diffusion Phenomenon by which gases mix together.

diode Electronic valve with two electrodes – an *anode* and a *cathode*.

direct current Electric current that always flows in the same direction. See also *alternating current*.

dispersion A system of particles suspended in another substance, solid, liquid or gas; also, the splitting of white light into the spectrum of colours.

distillation A technique for purifying or separating liquids by heating to boiling point and condensing the vapour produced back to a liquid.

Doppler effect The apparent change in frequency of sound or light caused by the movement of the source of the fre-

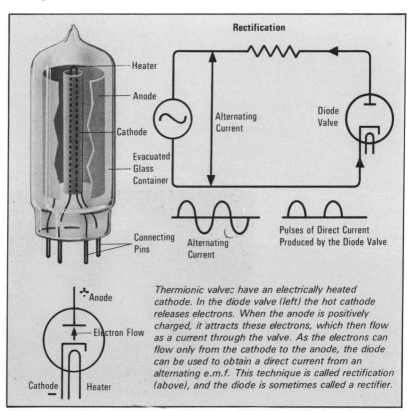

Thermionic valves have an electrically heated cathode. In the diode valve (left) the hot cathode releases electrons. When the anode is positively charged, it attracts these electrons, which then flow as a current through the valve. As the electrons can flow only from the cathode to the anode, the diode can be used to obtain a direct current from an alternating e.m.f. This technique is called rectification (above), and the diode is sometimes called a rectifier.

A-Z of Science

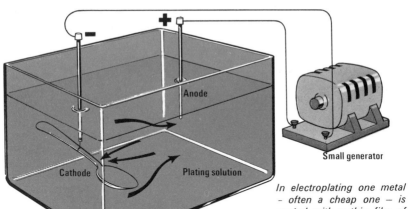

In electroplating one metal – often a cheap one – is coated with a thin film of another more expensive metal. In the diagram above the spoon is being chromium plated. The spoon is made the cathode. When electricity is passed through the solution of chromic acid, positive chromium ions (chromium atoms that have lost their electrons) are attracted towards the negative cathode where they pick up electrons and are deposited as a thin, even film on the spoon.

quency with respect to the observer, or vice versa. Frequency rises for approaching sources, falls for receding ones. The frequency or pitch of a train whistle, for example, seems higher as the train is approaching and lower as it moves away.

double decomposition Chemical reaction between two compounds in which they form two new compounds; e.g. AB reacts with CD to form AC and BD.

dry ice Solid carbon dioxide.

ductility Property of a metal that allows it to be drawn out into wire.

dynamo Device for converting mechanical energy into electrical energy; in simple form it consists of an armature rotating between the poles of a powerful electromagnet.

dyne Unit of force; imparts acceleration of 1 cm/sec/sec to mass of 1 gram; equivalent to one ten-thousandth of a *newton*.

earth In electric circuits, connection to piece of metal connected to the Earth.

efficiency Ratio of energy output of a machine to the energy input, generally expressed as a percentage. It can never be greater than 100 per cent.

effusion Movement of gas under pressure through small holes.

elasticity Property of a material that makes it go back to its original shape after a force deforming it is removed; if stressed beyond the *elastic limit*, the material does not return to its original shape. See also *Hooke's law*.

electric field Region surrounding electric charge in which a charged particle is subjected to a force.

electrical energy See *energy*.

electrode Metal plate through which electric current enters or leaves an electrolysis cell, battery, or vacuum tube. See *anode* and *cathode*.

electrolysis Conduction of electricity between two electrodes, through solution or molten mass (electrolyte) containing

205

ions; accompanied by chemical changes at the electrodes.

electromagnet Magnet consisting of an iron core surrounded by a coil of wire carrying an electric current. The core is magnetized only when the electric current is flowing.

electromagnetism Magnetism produced by a flowing electric current; the science that studies this phenomenon.

electromotive force (emf) The force that drives electric current along a conductor; measured in *volts*. See also *potential difference*.

electron Negatively charged atomic particle; every neutral atom has as many orbiting electrons as there are protons in the nucleus. See also *current, electric*.

electron microscope Instrument that uses a beam of electrons to produce magnified images of objects that are too small even to be seen with ordinary optical (light) microscopes.

electron-volt (eV) Unit of energy equal to the work done on an electron when it passes through a potential difference of 1 volt; $1 \text{ eV} = 1\cdot 6 \times 10^{-19}$ joule; i.e. 6 million million million electron volts equals one joule.

electroplating Production of a thin adherent coating of one metal on another by means of *electrolysis*.

element Substance made up entirely of exactly similar atoms (all with the same atomic number). The elements are listed, with their chemical symbols and atomic numbers, on p. 157.

emulsion Colloidal suspension of minute droplets of one liquid dispersed in another liquid. See *colloid*.

energy Capacity for doing work; examples include: *chemical energy* (possessed by a substance in its atoms or molecules, and released in a chemical reaction); *electrical energy* (associated with electric charges and their movements); *heat* (possessed by a body because of the motion of its atoms or molecules — a form of kinetic energy); *kinetic energy* (possessed by a body because of its motion); and *potential energy* (possessed by a body because of its position). In the presence of matter, any one form of energy can be converted into another.

erg Absolute unit of work in the cgs system, defined as the work done by a force of 1 dyne acting through 1 cm; $1 \text{ erg} = 10^{-7}$ joule; i.e. 10 million ergs equals 1 joule.

esters Class of organic compounds formed by the chemical reaction between acids and alcohols; e.g. ethyl acetate.

evaporation Phenomenon in which liquid turns into vapour without necessarily reaching the boiling point. It occurs because fast-moving molecules escape from the surface of the liquid. See also *boiling point*.

Fahrenheit Temperature scale on which the melting point of ice is 32°F and the boiling point of water is 212°F. To convert Fahrenheit to Celsius, subtract 32, multiply by 5 and divide by 9.

falling bodies, laws of Distance an object falls from rest under gravity in a vacuum is given by $s = \frac{1}{2}gt^2$, where s is distance, g is acceleration due to gravity, and t is time taken. Its final velocity is given by $v = u + gt$, where u is initial velocity. These equations are true for any body accelerating at a uniform rate.

fallout General term for radioactive fragments that fall to Earth from the atmosphere after a nuclear explosion.

farad (F) Unit of electrical *capacitance*, defined as that which requires a charge of 1 coulomb to raise its potential by 1 volt. It is the SI unit of capacitance. The practical unit is the *micro-farad*, which is a millionth of a farad (10^{-6}F).

fermentation Chemical reaction of organic substances brought about by living organisms such as yeasts and bacteria.

filtration Process for separating solids from suspension in a liquid by straining them off by means of special paper or another filter medium; the pure liquid that drips through is the *filtrate*.

fission (splitting) In atomic or nuclear fission, the nuclei of heavy atoms split and

A-Z of Science

ELECTROMAGNETS IN USE

Strong electromagnets are used in industry to lift heavy steel and iron weights, such as the scrap metal shown in the photograph.

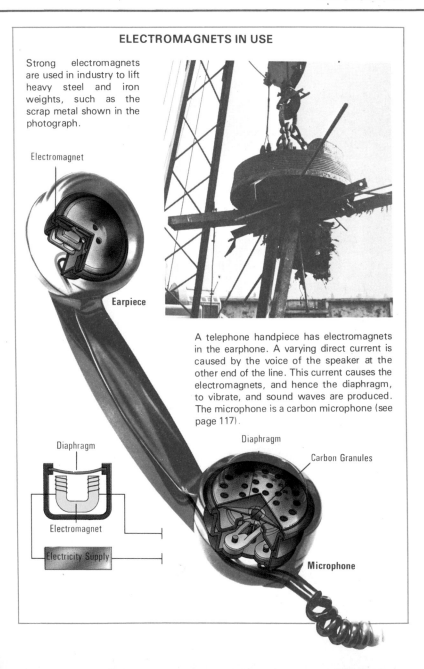

A telephone handpiece has electromagnets in the earphone. A varying direct current is caused by the voice of the speaker at the other end of the line. This current causes the electromagnets, and hence the diaphragm, to vibrate, and sound waves are produced. The microphone is a carbon microphone (see page 117).

A-Z of Science

release vast quantities of energy; this is the energy-generating process that takes place in atomic bombs and nuclear reactors.

flame test Technique in chemical analysis in which an element may be identified by the characteristic colour it imparts to a gas flame.

flexibility Property of a substance such as rubber or plastic that allows it to be bent easily without breaking. It is like *elasticity*, except that elastic materials return to their original position after being released.

fluid Substance (liquid or gas) that takes on the shape of part or all of the vessel containing it.

fluorescence Emission of light of one wavelength (colour) after absorption of another wavelength; it ceases when the light source is removed (unlike *phosphorescence*).

flux Substance added to help melting, or in soldering, to clean metal surfaces to be joined.

focus The point at which converging rays of light meet.

force Anything that can act on a stationary body and make it move, or make a moving body change speed or direction.

freezing point Temperature at which a liquid changes to a solid. It is the same as the *melting point* of the solid.

frequency Of a wave motion, the number of oscillations, cycles, vibrations, or waves per second; equal to 1 divided by the period (time taken for one vibration). See *wavelength*.

friction Force that resists sliding or rolling of one surface in contact with another.

fulcrum The point of support of a lever.

fusion, nuclear The joining of nuclei of light atoms together with the release of vast amounts of energy; this is the process that occurs in hydrogen bombs and in stars. Also known as a *thermonuclear reaction*.

gamma-rays (γ-rays) Penetrating electro-magnetic radiation of shorter wavelength than X-rays. They are emitted by the nuclei of radioactive atoms.

gas Substance that no matter how little there is always takes up the whole of the vessel containing it. See also *vapour*.

Geiger counter Instrument for detecting and measuring radioactivity.

Geissler tube Electric discharge tube that glows when a high voltage is applied between metal plates sealed into it; used for showing the effects of an electrical discharge through gases.

gravitation Force of attraction between any two objects because of their masses. *Newton's law of gravitation* states that the force between two particles is directly proportional to the product of their two masses, and inversely proportional to the square of the distance between them.

gravity The gravitational force between the Earth (or any other planet or moon) and an object on its surface or in its field of gravitation.

gravity, centre of Point at which all of an object's weight appears to act.

halogens Term for a group of elements — fluorine chlorine, bromine, and iodine — with closely related properties.

heavy water Deuterium oxide, D_2O; the oxide of deuterium, or 'heavy oxygen', an isotope of hydrogen having an atomic mass of 2.

henry (H) Unit of electrical inductance, defined as that producing an induced electromotive force of 1 volt for a current change of 1 ampere per second. It is the SI unit of inductance.

Hooke's law Within the elastic limit of a material, the extension (*strain*) is proportional to the force (*stress*) producing it. See *elasticity*.

hydrate Compound containing chemically combined water molecules; many salts are hydrated by water of crystallization.

hydrocarbon Chemical compound containing only hydrogen and carbon.

hydrolysis Decomposing a substance chemically by making it react with water.

hydroxide Chemical compound con-

A-Z of Science

A Geiger counter detects and records the number of radiation particles that pass through it. It contains a tube of gas with a central wire carrying a positive electric charge. As the particle passes through the gas, it knocks electrons from the atoms. The electrons travel to the wire and produce a signal that is recorded on a meter or as a click from a loudspeaker.

Right: A prospector uses a radiation detector to find the tell-tale radiation that radioactive minerals produce. The radiation of the minerals is very weak and harmless because the radioactive elements they contain are present in low concentrations.

taining the hydroxyl group —OH. It is derived from water by replacing one of the hydrogen atoms in the molecule with another, different atom or group.

ignition point The temperature above which a substance will burn.

incandescence The emission of light by a substance, caused by high temperatures.

inclined plane Simple machine consisting of smooth plane sloping upwards; used for moving heavy loads with relatively small force.

indicator Substance that changes colour to indicate the end of a chemical reaction or to indicate the pH (acidity or alkalinity) of a solution. See *pH*.

inductance Property of an electrical circuit to resist a changing current; in a single circuit it is called *self-inductance*; between two circuits, *mutual inductance*; symbol of inductance, L; the unit of inductance is the *henry*.

induction, electric In *electromagnetic induction*, a moving or changing magnetic field induces an electric current in a conductor. In *electrostatic induction*, a charge on a conductor induces an opposite charge on a nearby uncharged conductor.

induction coil High-voltage generator that makes use of electromagnetic induction between a coil of wire with few turns (primary) and a secondary coil with thousands of turns.

inert gases A group of chemically inactive gases – helium, neon, argon, krypton, zenon, radon; also called the noble, or rare, gases.

inertia Property of an object that makes it resist being moved or its motion being changed.

infra-red rays Electromagnetic radia-

A-Z of Science

tion of wavelengths just longer than those of visible light; invisible heat radiation.

insulation Use of an insulator, a substance that is a poor conductor of heat, sound, or electricity, to prevent their passing through.

interference Phenomenon caused by the superposition of waves; spectral colours produced in thin films and by diffraction are examples.

ionization Production of charged ions from electrically neutral atoms or molecular fragments; it is generally achieved by electrical or chemical processes.

ion Atom or group of atoms carrying an electrical charge; positively charged ions are called *cations*, negatively charged ions *anions*. During electrolysis, anions move towards the *anode* (positive electrode) and cations move towards the *cathode* (negative electrode).

isotope One of two or more forms of an

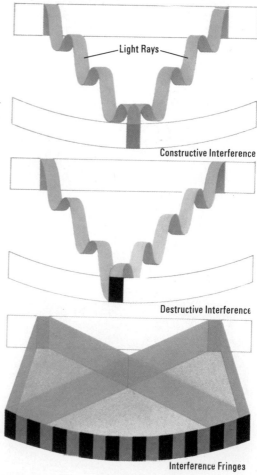

Right: Interference occurs when two light rays meet. If two identical rays meet so that the wave crests coincide, they will reinforce each other, producing a bright region on a screen (top). But if they meet so that the crests of one coincide with the troughs of the other, then they will cancel each other out, producing a dark region (centre). Identical rays are made by splitting a beam of light into two. Where the two resulting beams meet, a pattern of light and dark fringes is produced by interference (bottom). The pattern depends on the wavelength of the light and the distance each beam travels to the screen.

A-Z of Science

element with the same atomic number (i.e. number of protons in the nucleus), but different relative atomic masses (due to different numbers of neutrons in the nucleus). See also *atom*.

joule (J) Unit of work or energy defined as the work done when a force of 1 newton moves its point of application 1 metre in the direction of the force.

kelvin (K) Unit of temperature interval; the kelvin, or absolute, temperature scale is based on absolute zero as 0K; on this scale, the melting point of ice is 273·15K. To convert from kelvin to Celsius (or centigrade), subtract 273·15.

kinetic energy See *energy*.

laser Type of *maser* that produces an intense beam of light that is monochromatic (single colour) and coherent (all its waves are in step); abbreviation for *l*ight *a*mplification by *s*timulated *e*mission of *r*adiation.

latent heat Heat absorbed, without a rise in temperature, when a substance is changed from solid to liquid or liquid to gas.

lattice The regular network of fixed points in a crystal, about which atoms or molecules vibrate.

lens Device that affects light passing through it by converging (bringing together) or diverging (spreading apart) the rays.

lever Simple machine consisting of a

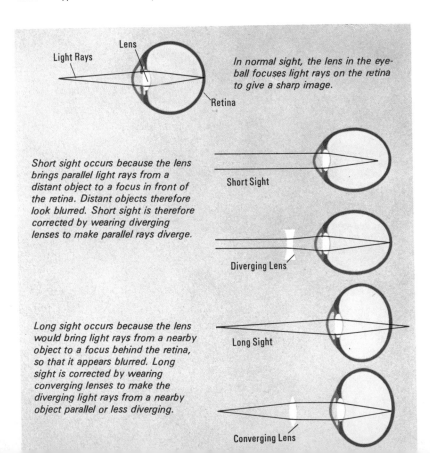

In normal sight, the lens in the eyeball focuses light rays on the retina to give a sharp image.

Short sight occurs because the lens brings parallel light rays from a distant object to a focus in front of the retina. Distant objects therefore look blurred. Short sight is therefore corrected by wearing diverging lenses to make parallel rays diverge.

Long sight occurs because the lens would bring light rays from a nearby object to a focus behind the retina, so that it appears blurred. Long sight is corrected by wearing converging lenses to make the diverging light rays from a nearby object parallel or less diverging.

A-Z of Science

rigid beam pivoted at one point, called the *fulcrum*; effort applied at one point on the beam can lift a load at another point.

liquid Substance that without changing its volume takes up the shape of all, or the lower part of, the vessel containing it.

litmus Vegetable dye used as a chemical indicator; it is red in acid solutions and blue in alkaline solutions.

longitudinal waves Waves in which the vibration or oscillation takes place in the same direction as the movement of the waves.

loudspeaker Device for converting electric currents into sounds that can be heard at a distance.

magneto High-voltage electric generator in which a permanent magnet is spun inside a coil; it gives only short pulses of current, and is used for starting petrol engines. See also *dynamo*.

malleability Property of a metal that allows it to be beaten into a thin sheet.

maser Microwave amplifier that uses energy changes within atoms or molecules; abbreviation for *m*icrowave *a*mplication by *s*timulated *e*mission of *r*adiation.

mass Amount of matter in an object; property of an object that gives it inertia. SI unit is the kilogram. See also *weight*.

melting point Temperature at which solid turns to liquid; equal to *freezing point* of the liquid.

metal Element or alloy that is a good conductor of heat and electricity, has a high density, often has a characteristic lustre, and can generally be worked by beating or drawing into wire.

microwaves Short-wavelength radio waves (electromagnetic radiation) having wavelengths of approximately 0·1 to 30 cm.

mixture More than one element or compound together, but not in chemical combination; components can be separated by physical means.

mole (mol) Basic unit of amount of a substance; it is defined as the amount that contains as many elementary entities as there are carbon atoms in 0·012 kg of carbon 12.

molecular weight Sum of the *atomic weights* of the elements in one molecule of a compound. Now called *relative molecular mass*.

molecule Smallest amount of a chemical substance that can exist alone; it is made up of one or more atoms.

momentum The product of the mass and velocity of a moving body. According to the principle of *conservation of momentum*, when two or more objects collide, the total of their separate momentums before impact equals the total momentum after impact.

motion, Newton's laws of (1) A stationary object remains still or a moving object continues to move in a straight line unless acted on by an external force. (2) The force producing acceleration in an object is proportional to the product of the object's mass and its acceleration. (3) Every action has an equal and opposite reaction.

Sir Isaac Newton

A-Z of Science

neutralization Combining acid and alkali so that the resulting solution is neutral (neither acid nor alkaline). See also *indicator*.

neutron Uncharged atomic particle found in the nuclei of all atoms except hydrogen.

newton (N) SI unit of force, defined as that imparting an acceleration of 1 metre per second per second to a mass of 1 kilogram.

noble metals Chemically unreactive metals such as gold, silver, and platinum.

nucleus, atomic The positively charged centre of an atom; consists of one or more protons and, except for hydrogen, one or more neutrons. See also *atom*.

ohm (Ω) SI unit of electrical resistance, given by the potential difference across a conductor in volts divided by the current in amperes flowing through the conductor.

Ohm's law A conductor obeys Ohm's law if the potential difference across it bears a constant ratio to the current flowing through it.

oscilloscope Electronic instrument in which an electric signal, or anything that can be reduced to an electric signal, is displayed as a 'trace' by a spot of light moving on the cathode-ray tube screen.

oxidation Making a substance combine with oxygen; removing hydrogen from a substance; or raising the oxidation number of an element (i.e. making it lose electrons). See *reduction*.

oxide Chemical compound of oxygen and another element.

ozone Form of oxygen containing three atoms in each molecule; O_3.

pascal (Pa) SI unit of pressure; 1 Pa = 1 newton per square metre.

Pascal's law In a fluid, the pressure applied at any point is transmitted equally throughout it.

pendulum Device consisting of a mass (bob) swinging at the end of a rigid or flexible support; *period* (time of one swing) of a simple pendulum, which has a flexible support such as a length of cord, is independent of the mass of the bob, depending only on the length of the cord and the acceleration due to gravity.

Periodic Table Organized arrangement of the chemical elements, in order of increasing atomic number; elements with similar chemical properties fall in vertical columns called *groups*.

pH Measure of acidity or alkalinity of a liquid; pH of 7 is neutral, lower numbers are acidic, higher numbers alkaline.

Below: The oscilloscope is an instrument for displaying electric signals in graph form. Varying electric fields deflect the fine electron beam emitted by the heated cathode, making it trace out a pattern on the fluorescent screen. The screen glows where the beam strikes it, thus producing a visible trace.

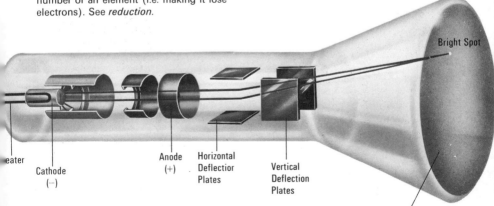

A-Z of Science

phosphorescence Re-emission of absorbed light even after the light source is removed. See *fluorescence*.

photon Quantum (packet of energy) of electromagnetic radiation, such as light.

polarized light Light in which electric and magnetic vibrations are restricted to two planes at right-angles, instead of being possible in all planes.

polymer A chemical compound with a high molecular weight, obtained by *polymerization*; most plastics are polymers.

porosity Property of a substance that allows gases or liquids to pass through it.

potential difference Defined as existing between two points of differing electric potential; if connected, current will flow between them; it is measured by the work done in moving unit charge between them. The SI unit of potential difference is the *volt*. See also *electromotive force*.

potential energy See *energy*.

power Rate of doing work, generally expressed in watts.

precipitation Formation and throwing out of solution of an insoluble compound as a *precipitate*; often achieved by *double decomposition*.

pressure Force per unit area; measured in newtons per square metre or dynes per square centimetre.

proton Positively charged atomic particle found in the nuclei of all atoms. The number of protons in an atom is the *atomic number*.

pulley Grooved wheel round which rope or chain runs; used for lifting weights.

quantum theory Theory that light and other forms of energy are given off as discrete packets (quanta) of energy.

radiation, heat Transfer of heat by means of waves; infra-red rays.

radioactivity Emission of radiation, such as *alpha-particles*, *beta-particles*, and *gamma-rays*, from unstable elements by the spontaneous splitting of their atomic nuclei.

reaction Chemical process involving two or more substances (*reactants*) and resulting in a chemical change.

reagent Chemical compound or solution used in carrying out chemical reactions.

reduction Making a substance react

A spectrum of colours is formed when a beam of light passes through a prism. In a spectrometer, the light first passes through a slit and then an image of the slit is formed by passing the beam through two lenses. In between the lenses, a prism splits up the light into its basic colours. A series of images of the slit in the different colours forms, overlapping to give the spectrum.

with hydrogen; removing oxygen from a substance; or lowering an element's oxidation number (i.e. making it accept electrons). See *oxidation*.

reflection Return or bouncing back of sound waves or electromagnetic radiation, such as a light ray, after it strikes a surface.

refraction Bending of a light ray as it crosses the boundary between two media of different optical density.

relative atomic mass. See *atom*.

relativity Einstein's theory that it is impossible to measure motion absolutely, but only within a given frame of reference.

relative density See *density, relative*.

resistance Property of an electrical conductor that makes it oppose the flow of current through it.

salt Chemical compound formed, with water, when a *base* reacts with an *acid*; a salt is also formed, often with the production of hydrogen, when a metal reacts with an acid. *Common salt* is sodium chloride.

saturated solution Solution that will take up no more of the dissolved substance.

semiconductor Substance, such as germanium and gallium arsenide, in which electrical resistance falls as its temperature rises. Used for making diodes and transistors.

SI units (Système International d'Unités)

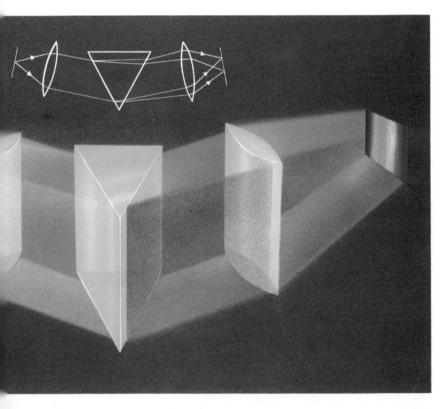

A-Z of Science

Internationally agreed system built round seven basic units, with several derived units, replacing other systems for scientific purposes.

solder Alloy with a low melting point, used for joining metals; alloys of tin and lead in varying proportions are used as soft solders; alloys of copper and zinc are used for brazing.

solenoid Coil consisting of many turns of wire wound round a hollow tube; it behaves as a magnet when carrying electric current.

solid State of matter that has a definite shape and resists having it changed; a crystalline solid melts to a liquid on heating above its melting point.

solubility Quantity of a substance (solute) that will dissolve in a solvent to form a solution. See *concentration*.

solvent Liquid part of a solution.

specific gravity Term now replaced by *relative density*. See *density, relative*.

specific heat capacity Quantity of heat needed to raise the temperature of unit mass of a substance by 1 degree; in SI units, joules per kilogram kelvin.

spectroscope Instrument for splitting the various wavelengths (colours) from a single light source into a spectrum, using a glass prism or diffraction grating.

speed Distance travelled by a moving object divided by the time taken. Speed in a particular direction is *velocity*.

static electricity Electricity involving charges at rest.

strain The proportion by which an object is deformed when a *stress* is applied. May be a ratio of lengths, areas, or volumes.

stress Force per unit area applied to an object.

sublimation Phenomenon in which a substance, on heating, changes directly from a solid to a gas or vapour without first melting to a liquid.

superconductivity Phenomenon, occurring at very low temperatures (approaching *absolute zero*), in which a metal continues to conduct an electric current without application of external electromotive force; the metal's resistance is effectively zero. The effect is used to produce large magnetic fields with the application of relatively small amounts of electrical energy.

surface tension Property of the surface of a liquid that makes it behave as though it were covered with a thin elastic skin; it is caused by the forces of attraction between molecules in the surface. See also *capillarity*.

suspension System consisting of very fine solid particles evenly dispersed in a liquid.

temperature Degree of hotness of an object referred to a selected zero (see *Celsius, Fahrenheit*) or to absolute zero (see *Kelvin*).

thermocouple Device for measuring temperature; consists of two different metals joined together, which generate a small electric current when the junction is heated.

thermometer Device for measuring temperature; common mercury thermometer is based on the expansion of mercury with rises in temperature, a thin thread of mercury rising up a tube as the temperature rises.

thermonuclear reaction See *fusion, nuclear*.

thermostat Control device, sensitive to changes in temperature, that maintains the temperature in an enclosure within a narrow, predetermined range.

transformer Electrical device used for converting alternating-current voltage to a higher or lower voltage.

transistor Semiconductor device that can amplify electric current. See *semiconductor.*

triode Electronic valve with three electrodes – an anode, a cathode, and a control grid.

UHF Ultra-high frequency radio waves.

ultrasonic waves 'Sound' waves beyond the range of human hearing.

ultraviolet rays Electromagnetic radiation of wavelengths just shorter than those

A-Z of Science

The behaviour of materials under stress must be known when a structure such as a suspension bridge is built.

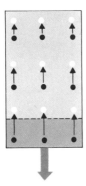

Tensile Stress stretches a body. The molecules or atoms in the lattice are pulled from their original position (white) to a new position (black), but the attracting forces between them oppose the force stretching the body. When the stretching force is removed, the attracting forces restore the body to its original shape. The strings of musical instruments vibrate in this way. This kind of elasticity is measured by Young's modulus of elasticity.

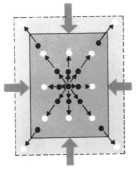

Shear Stress twists a body out of shape, endeavouring to divide it into layers. Attracting forces between the molecules restore it to its original shape. A diving board works by this kind of elasticity, which is measured by the shear modulus of elasticity.

Compressive Stress forces a body to reduce its volume. Repelling forces between the molecules restore its shape. Rubber heels on shoes and the cartilage between the bones of the spine make our lives more comfortable by undergoing this kind of elasticity, which is measured by the bulk modulus of elasticity.

A pond skater walks on the water, unable to penetrate the surface layer.

A-Z of Science

of visible light. Sun's radiation is rich in ultraviolet rays; they may be produced artificially with a mercury vapour lamp.

vacuum Ideally, region in which gas pressure is zero, i.e. a space in which there are no atoms or molecules; in practice, region in which pressure is considerably less than atmospheric pressure.

valency Number of hydrogen atoms (or their equivalent) with which an atom can combine; now obsolete.

vapour Gas that can be turned into a liquid by compressing it without cooling.

vector Physical quantity that needs direction as well as magnitude to define it (e.g. *velocity*).

velocity Rate of change of position; equal to *speed* in a particular direction.

VHF Very high frequency radio waves.

viscosity Property of a fluid (liquid or gas) that makes it resist internal movement; sticky or thick liquids are highly viscous, and as a result slow moving.

volt (V) Unit of electromotive force and potential difference, defined as potential difference between two points on a conductor if 1 joule of work is done when 1 coulomb of charge passes between them.

watt (W) Unit of electrical power, defined as the rate of work done in joules per second; equal to the product of current (A) in amperes and potential difference in volts (V). $W = AV$.

wavelengths Of a wave motion, the distance between crests (or troughs) of two consecutive waves; equal to the velocity of the wave divided by its frequency.

waves Regular disturbances that carry energy. Particles of a medium may vibrate; e.g. air molecules vibrate when sound waves pass, and water molecules vibrate when ripples cross water.

weight Force that an object exerts on whatever is supporting it. In gravitational units, weight is numerically equal to *mass*. The weight of an object on Earth differs from the weight of an object with the same mass on, say, the moon or another planet.

welding Joining together of two metal surfaces by heating them to a temperature sufficient to melt them and fuse them together.

work Done by a moving force; equal to the product of the force and the distance it moves along its line of action; measured in joules, ergs, etc.

X-rays Very short wavelength electromagnetic waves produced when a stream of high-energy electrons bombards matter.

A Van de Graaff generator is used to produce a high voltage. The bottom of the main column contains a set of spikes connected to an electricity supply of ten to fifty thousand volts. Electric charges are sprayed from the spikes on to a moving belt, and transferred to the dome at the top of the column (out of the picture). Here the charge can build up to several million volts. The dome of this generator has been connected to a steel ring, which is insulated from the ground. A steel girder placed near the ring causes the generator to discharge, and bright streams of 'lightning' flow between the ring and the girder

A-Z of Science

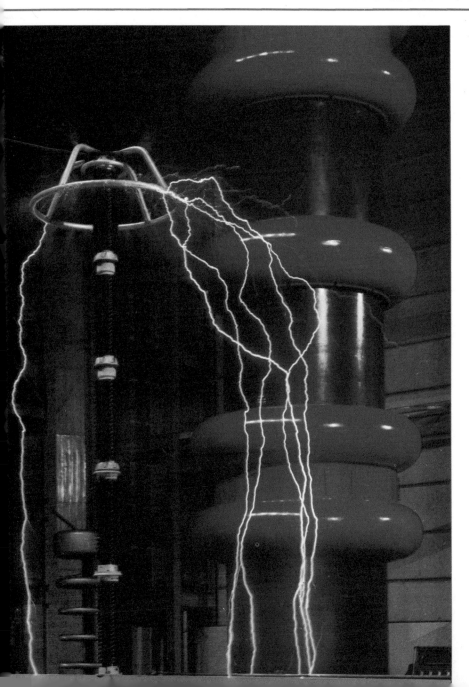

Famous Scientists

Avogadro Amadeo *(1776–1856),* an Italian physicist, is remembered mostly for his work on the atomic theory. In 1811 he suggested that equal volumes of gases at the same temperature and pressure contain the same number of molecules. It took 50 years before other scientists accepted this, and it later became known as Avogadro's law.

Bohr, Niels *(1885–1962),* a Danish physicist, did research on the structure of the atom at Cambridge under Sir J. J. Thomson and at Manchester under Lord Rutherford. Bohr added to our knowledge of the atom in many ways. He combined the quantum theory with atomic structure, and explained that electrons move around the nucleus in fixed orbits. Bohr suggested that the uranium atom could be split to release energy, as was later done in the atom bomb. He received the 1922 Nobel prize in physics.

Boyle, Robert *(1627–91),* an Anglo-Irish physicist and chemist, is often referred to as the 'father' of modern chemistry, because he took chemistry forward, away from alchemy. Boyle was especially noted for his experimental work, most of which he did in his laboratories at Oxford (1654–68) and London (1668–91). He invented a vacuum pump, which he used in 1662 to work on his theory on gases, now known as Boyle's law.

Curie, Marie *(1867–1934),* a Polish-born scientist who came to Paris to study physics and chemistry. In 1895 she married **Pierre Curie** and together they discovered the radioactive elements polonium and radium. For their work on radioactivity they shared the 1903 Nobel prize in physics with Henri Becquerel. Madame Curie became the first person to be awarded a second Nobel prize when, in 1911, she received the chemistry prize for

Sir Robert Boyle (1627-91)

The Curie family photographed in 1904: Marie, Pierre, and daughter Irene.

several elements, including potassium and sodium. He did extensive work with electricity and chemicals, but most schoolchildren remember him for the Davy lamp, which he designed to give miners a safe light and stop the explosions in mines. Davy was knighted in 1812, and elected president of the Royal Society in 1820.

Democritus (c.470–c.380 BC), a Greek philosopher, was known as the laughing philosopher, because of his happy nature. He is regarded as the greatest of all the Greek physical philosophers. Most of our present concepts of the nature of the world and of atoms were first suggested by Democritus. He believed that all things were made of atoms, which he said were tiny particles of the same matter but differed only in size, shape, and weight. He thought they were indivisible, indestructible, and constantly in motion. He explained the formation of the world by the separation of lighter and heavier atoms to form the Earth and the heavens.

Descartes, René (1596–1650), French philosopher and scientist, did most of his work in Holland, and became famous through his contributions to algebra and geometry and his attitude to philosophy and science. He devised the 'cartesian coordinate' system (a way to represent graphs) and introduced a new kind of thinking to the world of science.

Albert Einstein (1879-1955)

further work on radium. Her daughter Irene, together with her husband Frédérick Joliot, was also awarded the Nobel prize for chemistry in 1935. This was for their work on artificial radioactive elements. Irene Joliot-Curie died in 1956 of leukemia, the same disease that killed her mother 22 years earlier.

Dalton, John (1766–1844), an English scientist, did extensive work on the atomic theory. Dalton is best remembered for his famous Dalton's law of partial pressures, and for 'Daltonism' — a name for colour blindness, which he studied and from which he suffered. He taught mathematics and physics at Manchester. In 1825, for his work on the atomic theory, he received the medal of the Royal Society.

Davy, Sir Humphry (1778–1829), an English chemist and physicist, was woodcarver's son who became a professor of chemistry at the Royal Institution, London. He experimented with gases, and discovered the effects of nitrous oxide (laughing gas). He was the first to prepare

Einstein, Albert (1879–1955), an American physicist born in Germany, is regarded as one of the most brilliant physicists of all times. Most people know him for his famous theory of 'relativity'. During his early studies, he showed no signs of the genius that he was later to become. He lived and studied in Munich, Milan, and Switzerland, where he graduated and worked as an examiner in the patent office. There, some extra studies earned him a doctorate at the University of Zurich, where he formulated the Special Theory of Relativity and did other valuable work on many branches of physics. He later held many important posts in several universities, received the 1921 Nobel

Famous Scientists

Above: Suppose a man boards a train at 3 pm and travels at 90% of the velocity of light. After an hour has passed on the train and his watch says 4 pm, he looks out of the window at a passing station. The Special Theory predicts that everything would appear to be squashed to half its length and to weigh twice as much, and time to be slowed to half its normal rate. A man on the scales would appear to be very heavy and the station clock would read 3.30 pm (left). However, from the platform, the moving train would appear shortened in length, its mass doubled and its time slowed down (right).

Famous Scientists

physics prize, and was harassed by the German Nazis. In 1940 he became an American citizen, and continued his work, trying to establish a new theory, his 'unified field theory', which he never felt he had fully succeeded in proving.

Euclid, a Greek mathematician, lived and worked about 300 BC. Most of his work he recorded in a thirteen-volume book of geometry and other mathematics. Euclid's great contribution is his method of solving problems by stating the known facts and arguing with logical statements to the end.

Faraday, Michael *(1791–1867),* an English scientist, rose from being a bookbinder's apprentice to become a great man of science. He had little formal education, but his keen interest in science enabled him to progress rapidly from Sir Humphry Davy's assistant at the Royal Institution, London, to director of the laboratories there, and later to professor of chemistry. Faraday is best remembered for developing the first dynamo, discovering electromagnetic induction, and formulating 'Faraday's law'. He declined a knighthood as well as the presidency of the Royal Society.

Galileo *(Galileo Galilei) (1564–1642),* was an Italian astronomer, mathematician, and physicist. Much of modern experimental science developed from his investigation of natural laws. At the age of 19 he investigated and made important discoveries about the simple pendulum. Among his many posts, he was professor at the University of Pisa, where he started work on bodies in motion. He made an astronomi-

Below: Michael Faraday (1791-1867), went on to show that a magnetic field can be used to produce an electric current.

Below: From the General Theory of Relativity, Einstein predicted that stars would appear to change position when they are seen near the edge of the Sun. The Sun's gravity would cause light rays from the stars to curve as they pass near the Sun. The effect can only be seen during a solar eclipse, when the Sun's disc is obscured by the Moon.

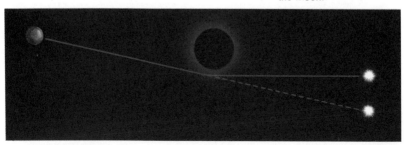

cal telescope, which is to this day used as a basis for some modern telescopes. The latter part of his life is sad, because his scientific discoveries led him to agree with Copernicus that the Sun is the centre of the Solar System. This view angered the churchmen of his time, who persecuted him and forced him to recant.

Galvani, Luigi (1737–98), was an Italian physician from whose name many terms in electricity, such as 'galvanizing' and 'galvanometer', are derived. Galvani was professor of anatomy at the University of Bologna. During one of his experiments on muscle and nerve, he noticed that a frog's leg twitched when touched with charged metal. Galvani made an arc of two metals, which also made the legs twitch, and wrongly explained it as electricity in the muscle. Volta disagreed and correctly attributed the electricity to the arc.

The Italian Galileo challenged many accepted theories and was condemned for his 'heretical' views.

Luigi Galvani (1737-1798) was born in Bologna. He later studied and practised medicine there. As a result of his experiments he thought he had discovered that electricity could be obtained from the body of an animal. He was wrong; he had in fact discovered the electric cell.

Joule, James Prescott (1818–89), an English physicist, was one of the great experimental scientists of the 1800s. The electrical unit of work is named after him. Joule invented an electromagnetic engine as a youth, and he later did valuable work on heat, thermodynamics, and electricity. His work on heat established the relationship between heat energy and mechanical energy, but his greatest contribution is his discovery of the 'first law of thermodynamics'.

Lavoisier, Antoine Laurent (1743–94), a French chemist and physicist, is said to be the founder of modern chemistry. He was able to extend and co-ordinate the work of others because of his clear logical thought and ability to experiment and interpret his results. He explained what combustion is, and so disproved the theory that a substance called phlogiston was given off during burning. Lavoisier held many government jobs, and was guillotined in the French Revolution.

Famous Scientists

The Russian chemist Dmitri Ivanovich Mendeleyev, whose work laid the foundation for the modern periodic classification of the chemical elements. Born in 1834 in Tobolsk, Siberia, Mendeleyev studied at St Petersburg (now Leningrad) and at Heidelburg, Germany. He wrote a classic chemical textbook Principles of Chemistry between 1868 and 1870. He died at St Petersburg in 1907, by which time the gaps he left in his table had been filled, as he had predicted, by new elements.

Lomonosov, Mikhail Vasilievich *(1711–65), a Russian scientist and writer, is regarded as the 'father' of Russian science. He studied the sciences and philosophy in Germany, and in 1741 received a lifetime appointment to the Russian Academy of Sciences. It is clear from his experiments, though it is not known whether by chance or by design, that his thoughts were in agreement with such modern principles as the kinetic theory of gases and the mechanical nature of heat.*
Maxwell, James Clerk *(1831–79), a Scottish physicist, revolutionized fundamental physics. He was professor at Aberdeen, London, and Cambridge. Besides his work on mathematics and several other topics, he is remembered especially for his theory of the electromagnetic field, which led to the most important 'Maxwell's equations'. This work supported the wave nature of light, and classified it as another form of electromagnetic radiation.*
Mendeleyev, Dmitri Ivanovich *(1834– 1907), a Russian chemist, is famous for*

the periodic classification of the elements in the Periodic Table. This arrangement placed elements with similar properties in groups (vertical columns), and periods (horizontal rows), in a cyclic table. With it, Mendeleev was able to predict the properties of some elements before they were discovered.

Newton, Sir Isaac (1642–1727), an English philosopher and physicist, ranks among the greatest scientists that ever lived. He studied at Cambridge, where he became professor of mathematics. He made many important discoveries did experiments, and formulated theories — most important of which are the law of universal gravitation, the calculus, the spectrum of colours of white light, his theory that light is composed of corpuscles (particles), and his three laws of motion. The wave theory of light has been combined with the corpuscular theory in the modern quantum theory. Newton and Leibniz are jointly credited with having invented the calculus – a cause for dispute between both men.

Sir Isaac Newton, regarded as one of the most brilliant of all scientists.

Nobel, Alfred Bernhard (1833–96), a Swedish chemist educated in Russia, was the son of an explosives manufacturer. He helped his father develop torpedoes, mines, and other explosives. After a series of fatal accidents in the family factory, Nobel developed dynamite as a safer explosive. He had misgivings about his work, and left a fund for raising money to be awarded annually as the Nobel prizes for work in physics, chemistry, physiology, medicine, and literature, and for the promotion of international peace.

Oersted, Hans Christian (1777–1851), a Danish physicist and chemist, had the unit of magnetic field strength named after him. The study of electromagnetism was prompted by his discovery in 1819 that an electric current in a conductor deflects a magnetic needle.

Pascal, Blaise (1623–62), was a French scientist and religious philosopher, whose father, a civil servant, directed his studies. At an early age, Pascal showed signs of being a brilliant scientist. He wrote a paper of good quality on conic sections before he was 16, and invented a calculating machine at 19. He formulated the modern theory of probability, and proposed 'Pascal's law', which applies to pressure in fluids. When his father died and he himself narrowly escaped death, Pascal turned to religion.

Pauling, Linus Carl (born 1901), an American chemist, won two Nobel prizes – one for chemistry (1954) and the other for peace (1962). Pauling's use of the 'quantum theory' in calculations of molecular structure was a new idea. Probably his most important work is his explanation of covalent bonds. He also made contributions in molecular biology, recommended the use of large quantities of vitamin C for treating the common cold, and suggested the use of chemicals to cure mental diseases. He was awarded his peace prize for work towards nuclear disarmament.

Planck, Max (1858–1947), a German physicist, laid the foundations for the quantum theory. In 1900, he advanced the theory that atoms emit and absorb energy in packets called quanta, instead of continuously as was thought. This prepared the way for Einstein, Bohr, and others to arrive at the quantum theory. Planck is remembered also for his work on 'black body' radiation, and for 'Planck's constant'. He was president of the Kaiser Wilhelm Society which, after World War II, became part of the Max Planck Institute. He received the Nobel prize (1918) in physics.

Priestley, Joseph (1733–1804), an English theologian and scientist, supported the view that combustion produced a substance called phlogiston, so that when he discovered what Lavoisier later named oxygen, he called it 'dephlogisticated air'. Priestley did many valuable scientific experiments, some of which led him to discover gases such as sulphur dioxide and ammonia. He explained the rings formed by an electric discharge on metal, now called Priestley's rings.

Pythagoras (c.582–c.507 BC), a Greek philosopher, founded a secret religious society. Little is known of his life and writings, because they are indistinguishable from those of his disciples. He is said to be the first to use the word 'philosophy'. His society – the Pythagoreans – discovered that the relationship between musical notes could be explained by mathematics, and went on to devise a theory of numbers. Pythagoras is best remembered for his theorem of right-angled triangles.

Rutherford, Ernest, Baron Rutherford of Nelson (1871–1937), British physicist born in New Zealand. He is known, apart from his work on the atom and radioactivity, for his leadership in directing the research of others. In 1919 he succeeded J. J. Thomson, his former professor, as professor and director of the Cavendish Laboratory at Cambridge. He worked extensively on radioactivity, and discovered the nucleus, alpha-particles, and beta-particles. He supervised work which led him to explain the atomic structure, and provided first evidence of artificial splitting of the nucleus. He received the 1908 Nobel prize in chemistry.

Volta, Count Alessandro (1745–1827), an Italian physicist, famous for his work on electricity. He made many inventions, including the 'voltaic pile', the electrophorus, an electric condenser, and the 'voltaic cell'. He was professor of physics at the University of Pavia, and was made a count and a senator by Napoleon I. The unit of potential difference, the volt, is named after him.

Below: Ernest Rutherford, famous for his work in the field of atomic physics.

BRANCHES OF SCIENCE

Science	Study of
Earth Sciences	
Geology	rocks, earthquakes, volcanoes, and fossils
Meteorology	the atmosphere and weather
Mineralogy	minerals, their location and mining
Oceanography	waves, tides, currents, trenches, and ocean life
Palaeontology	plant and animal fossils
Petrology	formation and structure of rocks; their chemical content
Life Sciences	
Agronomy	land management of crops and cultivation
Anatomy	structure, form, and arrangement of the body
Bacteriology	bacteria, their growth and behaviour
Biology	animals and plants; origin, morphology, and environment
Botany	the plant world
Cytology	structure, function, and life of cells
Ecology	relationship between living things and environment
Medicine	cause, prevention, and cure of disease
Nutrition	supply of adequate and correct foods to satisfy the body's requirements
Pharmacology	drugs; their preparation, use, and effects
Physiology	the function of living things
Psychology	behaviour of humans and animals; working of the brain
Zoology	animals
Mathematical Sciences	
Logic	reasoning by mathematics; used by computers
Mathematics	the application of geometry, algebra, and arithmetic, etc.; application of these to concrete data
Statistics	numerical information which is to be analysed
Physical Sciences	
Aerodynamics	the properties and forces of flowing air on solid objects
Astronomy	heavenly bodies and their motions
Chemistry	properties and behaviour of substances
Electronics	behaviour of electrons in a vacuum, in gases, and semi-conductors
Engineering	application of scientific principles to industry
Mechanics	the invention and construction of machines, their operation, and the calculation of their efficiency
Metallurgy	the working of metals; smelting and refining
Physics	nature and behaviour of matter and energy
Social Sciences	
Anthropology	origin, culture, development, and distribution of man
Archaeology	remains, monuments left by prehistoric man
Economics	use of natural resources to the best advantage
Geography	location of Earth's features and man's relation to them
Linguistics	languages and their relationship to each other
Political Science	function of states and governments
Sociology	Relationship between groups and individuals

SI UNITS

Basic units

	Symbol	Measurement
metre	m	length
kilogram	kg	mass
second	s	time
ampere	A	electric current
kelvin	K	thermodynamic temperature
mole	mol	amount of substance
candela	cd	luminous intensity

Derived units

hertz	Hz	frequency
newton	N	force
pascal	Pa	pressure, stress
joule	J	energy, work, quantity of heat
watt	W	power, radiant flux
coulomb	C	electric charge, quantity of electricity
volt	V	electric potential, potential difference, emf
farad	F	capacitance
ohm	Ω	electric resistance
siemens	S	conductance
weber	Wb	magnetic flux
tesla	T	magnetic flux density
henry	H	inductance
lumen	lm	luminous flux
lux	lx	illuminance

Supplementary units

radian	rad	plane angle
steradian	sr	solid angle

USEFUL FORMULAE

Circumference
Circle $\quad 2\pi r$

Area
Circle	πr^2
Surface of sphere	$4\pi r^2$
Ellipse, semi-axes a, b	πab
Triangle, base b	$\frac{1}{2}bh$
Rectangle, sides a, b	ab
Trapezium, parallel sides a, c	$\frac{1}{2}h(a+c)$
Regular pentagon, side a	$1 \cdot 721a^2$
Regular hexagon, side a	$2 \cdot 598a^2$
Regular octagon, side a	$4 \cdot 828a^2$

Volume
Sphere	$\frac{4}{3}\pi r^3$
Cylinder	$h\pi r^2$
Cone	$\frac{1}{3}h\pi r^2$
Rectangular prism, sides a, b, c	abc
Pyramid, base area b	$\frac{1}{3}hb$

$r =$ radius, $h =$ height

algebraic
$a^2 - b^2 = (a+b)(a-b)$
$a^2 + 2ab + b^2 = (a+b)^2$
$a^2 - 2ab + b^2 = (a-b)^2$
For quadratic equation $ax^2 + bx + c = 0$

$$x = \frac{-b \pm \sqrt{b^2 - 4ac}}{2a}$$

DECIMAL MULTIPLES

prefix	symbol		multiplication factor
tera	T	10^{12}	1,000,000,000,000
giga	G	10^9	1,000,000,000
mega	M	10^6	1,000,000
kilo	k	10^3	1,000
hecto	h	10^2	100
deca	da	10	10
deci	d	10^{-1}	0·1
centi	c	10^{-2}	0·01
milli	m	10^{-3}	0·001
micro	μ	10^{-6}	0·000001
nano	n	10^{-9}	0·000000001
pico	p	10^{-12}	0·000000000001
femto	f	10^{-15}	0·000000000000001
atto	a	10^{-18}	0·000000000000000001

INDEX

Page numbers in *italics* refer to illustrations.

A

Abacus 96, *97*
Absolute scale *see* Kelvin scale
Absolute zero 112, 200
Absorption 200
Absorption spectrum *42*
a.c. *see* Alternating current
Acceleration 34, 41, 200
Accelerator 15
Accumulator 200
a.c. electric motor 78–81, *78–81*
Acids 130, 200
Acrylics 140
Acoustics 105
Actinide series *199*
Actinium 142
Adsorption 200
Aerial 58, *59*, 60, *61*, *63*, *85*, 86
Aerodynamics 228
Agronomy 228
Aiken, Howard 97
Air 112, *114*, 117, *123*
Aircraft 33, *34*, *105*
Airship 197
Alkali metals 178, *179*
Alkalis 130, 200
Alloy 200
Alpha particles *12*, *13*, 14, 16, 200
Alternating current 78, *80*, *82*, 83, *83*, 84, 87, 200
Alternator 83
Aluminium 116, 142, *143*
Americium 142
Ammeter *76*, 200
Ammonia 136, 137, 192, *192*, 193, *193*
Ampere 69
Amplifier 60, *107*, 201
Amplitude *44*, *56*, *58*, 60, *86*, 88, 89, 200, *216*
Analog computer 98
Analytical chemistry 136
Anatomy 228

Angstrom unit 200
Angular momentum 38
Anhydrous 201
Annealing 201
Annealing furnace *159*
Anode *84*, 200
Antenna *see* Aerial
Anthropology 228
Antimony 142, 175
Applied chemistry 136
Archaeology 228
Archimedes principle 200
Area 229
Argentite 170
Argon 142, 190, *191*
Armature 76, *77*, 78, 200
Arsenic 142
Asbestos 180
Assembly line *125*
Astatine 142
Astronomy 228
Atmosphere 182–185, *183–185*
Atom 11–23, *10–23*, 200
Atomic clock 178
Atomic number *198*, 200
Atomic particles 41
Atomic weight *198*, 201
'Atom smasher' *see* Particle accelerator
Aurora *183*
Average value 83

B

Babbage, Charles 96
Bacteriology 228
Baird, John Logie 90, *92*, 95
Barium *151*
Base *89*
Bakelite 130, 140
Balance *132*, *135*
Barium 180, 181
Barium meal *181*
Barytes 181
Base 201
Basic-oxygen furnace 162, 163
Battery, electric 66, *66*, 67, 201
Beaker *135*
Beats *109*
Becquerel, Henri *13*
Berkelium 142
Beryllium 142, 180

Beryllium-polonium core *20*
Bessemer, Henry 161
Beta particle 12, *13*, 14, 201
Bimetallic strip *116*
Binary code *96*, 98
Biochemistry 136
Biology 228
Bismuth 142
Black *54*, 56, 57
Blast furnace *158*, 159
Blue *54*, 55
Bog-iron ore *see* Limonite
Boiling point 112, 113, 118, *118*, 119, 201
Bomb *20*, 21
Bonding *14*, *15*
Boron 143
Botany 228
Boyle's law 201
Brass 166
Breeder reactor *see* Fast reactor
Broadcasting 89, 91, 93
Bronze 166, *166*, 175
Bronze Age *166*
Brush 73, *74*, *75*, 89, 81, *81*
Bunsen burner *135*, *138*, *186*, 188
Burette *135*

C

Cadmium 18
Caesium 144, 178
Calcium 144, 180, 181
Calculating machine 96
Calendering 140
Californium 144
Calorie 201
Camera, television 60, 62, *90*, *91*, *94*
Capacitance 201
Capacitor (condenser) *69*, *88*, *89*, 201
Carat 172
Carbon 144
Carbon dioxide 182, 186, 189
Carbon monoxide 189
Carothers, W. H. 141
Carnallite 180
Carrier waves *56*, *58*, 60, *86*, *216*
Cash register *98*
Cassette recorder 108
Cassiterite 175

Cast iron 160
Catalyst 134, *134*, 172, *193*, 201
Cathode *84*, *90*, 201
Cathode-rays *62*, 91, *91*, 201
Caustic soda 136, *137*, *138*
Cell, electric 67, *67*, *69*, *70*
Celluloid 130, 138, 140
Cellulose 138, *139*
Cellulose acetate 140
Celsius, Anders 112
Centigrade scale (Celsius) 112, 116, 204
Centrifugal and centripetal forces 37, 201
Centrifuge *135*
Chain reaction 16
Chalcopyrite 168
Charles' law 201
Chemical elements 142–152
Chemical energy *27*, 28, *29*
Chemicals 130–140, *130–140*
Chemical symbols *198*
Chemistry 228
Chile saltpetre 193
Chinook winds *123*
Chlorine *137*, 145
Chromium 145
Circuits see d.c. circuits; Computers
Circular motion 37, *37*
Clinical thermometer *112*
Clouds *194*
Coal tar 130
Cobalt oxide 145
Coherent light 51
Coherer 84
Coinage metal 168, 169, 175
Coke 160
Cold 110, 116, 128, *129*
Collector *89*
Collision *30*, *31*
Colloidal state 202
Colour 54–56, *54*, *55*
Colour mixing 56
Colour television 62, 93–95, *93–95*
Combustion 188
Communications satellite 59
Commutator 73, 74, *74*, *75*, 81, *81*
Compass 38, 71
Compression stroke *124*
Computers 96–101, *96–101*
Concave 46, *47*, 48, 202

Condensation 120, 202
Conductor, electrical *65*, 66, 69, 171, 178, 202
Conduction, thermal 115, 118, 202
Control rod 18, *18*, *19*
Convection 114, 117, 202
Converging lens *45*, *47*, 48
Convex 45, 46, *47*, 48, 202
Coolant 18, *19*
Copper 114, 145, 164–166, *164–166*
Copper oxide 165
Core, the 18, *18*, *19*
Corrosion 202
Cosmic rays 44, *183*, 202
Counting 96
Couple 37
Covalent bond *14*
Cracking 134
Critical mass 17
Cryogenics 129
Crystal 202
Crystal diode 87
Crystal pick-up *107*
Cupro-nickel 166
Curie, Marie and Pierre *13*
Curium 145
Current see Alternating current; d.c. Circuit
Current electricity 64
Cyan *54*, *55*, 57
Cycles: nitrogen *192*; water *194*
Cyclotron 202
Cytology 228

D

Dacron 141
d.c. see Direct current
d.c. electric motor 73–77, *73–77*
Decimal multiples 229
de Forest, Lee 87
Demodulator 60, *61*, 89
Density 202
Deuterium *20*, 21, 23
Dichroic mirror *94*
Dielectric 204 see also Insulator
Diffraction 204
Diffusion of gases 204
Digital computer 96
Diode *85*, 86, 204

Direct current 68, 73, 204
Direction 35, *35*
Disc recording 106, 108
Dispersion 204
Distillation 134, 186, *188*, 204
Diverging lens 46, *47*, 48
Dolomite 180
Doppler effect 204
Double glazing 117
Downs cell *179*
Dry cell 67
Ductile 205
Dyes 130
Dynamite 130
Dynamo effect *73*
Dyne 205
Dysprosium 145

E

Earth sciences 228
Eckert, J. Presper 97
Ecology 228
Economics 228
Edison, Thomas Alva 86, *106*, 108
Edison effect 86
Einstein, Albert 17, 40
Einsteinium 145
Elasticity 205
Electrical energy 28, *29*
Electrical-arc furnace 162, 163
Electric bell 72
Electric field *44*, *78*, 79, 205
Electricity 64–70, 64–70, 83 see also Electromagnetism
Electric light bulb *191*
Electric wind *65*
Electrode 67, *163*
Electrolysis 134, 165, 205
Electrolyte 67, 205
Electromagnet *70*, 72, 76, *109*, 206
Electromagnetic induction 72
Electromagnetic interaction 25
Electromagnetic radiation 13, 43, 58, 84
Electromagnetism *68*, 71–81, *70–81*
Electron 11, *12*, *13*, 16, 206
Electronic Numerical Integrator and Calculator

231

97
Electronics 228
Electron microscope 2
Electroplating 167, 173
Elements 11, 14, 206; periodic table *198–199*
Emission spectrum *42*
Emitter *89*
Encoder *63*
Energy 17, 26–29, *26–29*, 41, 206 *see also* Electricity; Heat
Engine *see* Jet engine; Petrol engine; Rocket engine; Steam engine
Engine block *110*
Engineering 228
ENIAC *see* Electronic Numerical Integrator and Calculator
Erbium 145
Erg 206
Europium 145
Evaporation 119, 128, 194, *194*, 206
Exhaust stroke *125*
Exosphere *183*, 184, 185
Exothermic reaction 188
Expansion 31
Explosives 137
Extrusion 140

F

Fahrenheit, Gabriel 112
Fahrenheit scale 112, 116, 206
Fall-out 21
Faraday, Michael 57
Far sight *see* Hypermetropia
Fast reactor 18, *18*
Fennec fox *115*
Fermi, Enrico *16*
Fermium 145
Fibreglass 141
Fibre Optics *46*
Film *169*
Filter 55, *55*
Fireplace 118
Fireworks *151*, 181
Fission 16, 17, 18, *19*, 206
Fission bomb *20*, 21
Flame *138*, *186*, 188
Flask *135*
Fleming, John Ambrose 86

Fluids 208
Fluorine 145
Focus, principal 208
Föhn winds *123*
Force 24–25, *24–25*, 30, *34*, *37*, 208
Ford, Henry *125*
Four-stroke cycle *124*, 125
Fractional distillation 186, *188*
Francium 146
Freezing 112, 113, 119, 208
Frequency 83, *88*, *89*, 208
Frequency modulation 44, *56*, *58*, 60, *86*, 89
Friction 33, *33*, *34*, 208
Fuel rod *18*, *19*
Fulcrum 208
Funnel *135*
Furnace *157*, *158*, 159, *159*, 162
Fusion 208
Fusion bomb *20*, 21
Fusion control 22

G

g 211
Gadolinium 146
Galena 175, *175*
Gallium 146
Galvanizing 211
Galvanometer *69*, *70*, *85*, 87
Gamma radiation 18
Gamma rays 13, 43, 211
Gangue 158
Gas *11*, 66, 128, 190, *190*, *191*
Gas matter 208
Geiger-Müller counter 209
Generator 66, 79, *79*, *80*, *81*
Geography 228
Geology 228
Geranium 147
Gold *146*, 147, *147*, 169, 170
Gold leaf 169
Gold Rush 170
Goonhilly satellite *59*
Gravity *24*, 25, *25*, 32, 34, *37*, 39, *39*, 41, 182, 208
Grid *84*
Ground waves *87*
Gun metal 166
Gyrocompass 38
Gyroscope 38, *38*, 52

H

Haematite 157
Hafnium 147
Hahnium 147
Headphones *86*, 89
Heat 110–127, *110–127*
Heat engine 122
Heat exchanger *19*, *123*
Heavy spar *see* Barytes
Hydrogen *14*, 22, 147, 191, 193, *193*, 195, 196, *196*, *197*
Helium *12*, *13*, 129, 147, 190, 191
Hero's engine *116*
Hertz 83
Hertz, Heinrich 57, 84, *84*
Hiroshima 21
Holmium 147
Holography 53
House heating 128
Hovercraft, tracked 77
Huygens, Christiaan 45, *45*
Hydrocarbons 130
Hydroelectric power *82*
Hydrogen bomb *20*, 21
Hydrogen bond 195
Hydrogen peroxide 196
Hydrolysis 208

I

Ice *118*, 120, 194, 195
Ice skating *119*
Image-orthicon camera tube *63*, *90*, 92, *94*
Indium 147
Inductance *88*, 209
Induction coil *84*, 209
Induction motor 77, 78, *78*
Induction stroke *124*
Inert gases 190, *190*, *191*, *198*, 209
Inertia 32, 209
Infra-red rays 43, 185, 209
Inorganic chemistry 136
Insulator 66, 210
Insulation 116
Integrated circuit 98
Interference 210
Interference fringes 53
Interferometer 52
Internal combustion engine 124, 125

Invar 168
Iodine 148
Ion 66, *67*, 185, 210
Ion engine *23*
Ionic bond *15*
Ionization 210
Ionosphere *63*, 87, *183*, 184
Iridium 170, 172
Iron 148, 154–163, *154–163*, *198*
Iron Age 154
Iron Knob, S. Australia *154*
Iron oxide 165
Iron pyrites 157
Isothermat 213
Isotopes 12, 14, 16, 210

J

Jet engine 33, 125, *126*
Jet streams 184
Joinery *117*
Joule 28, 114, 115, 211
Joule, James Prescott 28
Jupiter *62*

K

Karat *see* Carat
Kelvin 113, 114, 116
Kelvin, Lord 112
Kelvin scale 112
Kidney iron ore *see* Haematite
Kieserite 180
Kinetic energy *27*, 28, *29*
Kinetic theory of gases 213
Krypton 148, 190

L

Laboratory *131*, *135*
Lanthanide series *198*
Lanthanum 148
Lasers 23, 50–53, *51–52*
Latent heat *118*, 119, 128, 211
Lattice 211
Lawrencium 148
Lead 18, 148, 174
Lead glance *see* Galena
Length 40
Lens *45*, 46, *47*, 213
Lenz's law *73*, *80*

Lever 211
Life sciences 228
Light 40, *40*, *42*, 43–47, 44–47 *see also* Lasers
Light bulb 68, *191*
Light meter *70*
Lightning conductor *65*
Limonite 157, 158
Linguistics 228
Liquid air 129, *129*
Liquid-fuel rocket *126*, 127
Liquids *11*, 66, 116, 117, 118, 129
Linear induction motor 77
Lithium *20*, 21, 178
Litmus 212
Logic 228
Longitudinal wave 212
Long sight *see* Hypermetropia; Presbyopia
Loom *173*
Loudspeaker *72*, *86*, 89, 106, *107*
Lucite *see* Perspex
Lutetium 148

M

Magenta *54*, *55*, 57
Magnesite 180
Magnesium 148, 180, *180*
Magnet 73, 79, *80*, 81
Magnetic field *44*, 57, *68*, 71, *74*, *75*, *76*, 78, *80*, 82
Magnetic poles 68, *75*
Magnetic tape 108
Magnetism *68*
Magnetite 157
Magnification 46, *47*
Malleable 212
Marconi, Guglielmo 57, 85
Margarine *197*
Mass 17, 32, 40, 212
Mass spectrometer *130*
Mathematical sciences 228
Matte 165
Mauchly, John 97
Maxwell, James Clerk 57, 84
Measurement 229
Mechanical energy 28, *29*, 122
Mechanics 228
Medicine 228
Melting point *118*, 119, 120, 123, 212
Mendelevium 149

Mercury 149
Mesosphere 184
Metal 116, *116*, *117*, 215; alkali 178, *179*; alkaline earth 180, *180*, *181*; precious 169–173, *168–173*; refractory 176, *177*; symbols, weights and numbers 198–199 *see also* Tin; Lead; Zinc
Metallurgy 228
Meteorology 228
Methane 196
Microanalysis 136
Microphone *56*, *72*, *73*, 106, *108*
Microprocessor 101
Microscope *135*
Microwave *63*, 212
Mineralogy 228
Mining *155*, 170, *170*
Mirror 45, *45*, *47*, *94*
Moderator *19*
Modulation 52, 60, *86*
Molecule 212
Molybdenum 149, 176
Momentum *31*, 34, 38
Mone metal 166, 168
Mono recording 106
Morse code *85*
Mortar and pestle *135*
Motion 30–39, 40, *40*, *71*, 212
Motor effect *71*, *72*
Motor, electric 73–81 *74–81*
Moving coil instrument *72*, *73*, *76*
Muntz metal 166

N

Nagasaki 21
Natural gas 190, 191, 196
Neap tide *25*
Nebula *196*
Neodymium 149
Neon 182, 190
Neon lighting 66, 149
Neptunium 149
Neutralization 213
Neutron 11, 14, 213
Neutron radiation 18
Newcomen, Thomas 122
Newton 32
Newton, Isaac 30, 32, 34, 40

Nickel 149
Nickel 167, *167*
Nickel silver 166
Niobium 176
Nipkow, Paul 90
Nitrates *192*
Nitric acid 137, *192*
Nitrogen 149, 185, 186, *188*, *192*, *192*, *193*
Nobel, Alfred 130
Nobelium 149
Noble gases 190
Noble metals 169
Noise 104
Nuclear bombardment 14
Nuclear energy 28, 52
Nuclear physics 217
Nuclear reactor 17, *18–19*, *23*, *27*
Nuclear rocket engine *23*
Nuclear submarine 18
Nucleus, atomic 11, *12*, 16, 24, 213
Nutrition 228
Nylon 141

O

Oceanography 228
Ohm, Georg 70
Ohm's law 69
Open-hearth furnace *158*, 162
Organic chemistry 136
Oscillator 88
Osmium 149, 172
Overtones *see* Harmonics
Oxidation 134, 213
Oxyacetylene welding *187*
Oxygen *14*, 149, 182, 185, 186–189, *186–188*
Ozone 184, 185

P

Paint 56
Palaeontology 228
Palladium 149, 172
Parabola 37
Parsons, Charles 124
Particle accelerator 15
Particles *see* Radiation
Pascal, Blaise 96
P.D. *see* Potential difference

Pelletization 159
Pendulum 39
Pentlandite 168
Periodic table 198–199
Periodic time *82*, 83
Perspex 140
Petrol engine 122, 124, *124*, *125*, 189
Petroleum 130, 134
Petrology 228
Pewter 175
Pharmacology 228
Phonograph *106*, 108
Phosphor dots *95*
Phosphorescence 218
Phosphorus *12*, 149
Photoelectric cell *70*, 90, 92
Photography, colour 57
Photon 214
Photosynthesis 186
Physical chemistry 136
Physical sciences 228
Physics 228
Physiology 228
Pick-up *107*
Picture tube 93
Piezo-electricity *107*
Pig iron 160, 162
Pigment 56
Pipette *135*
Piston engine 124
Piston linings *110*
Planet 31
Platinum resistance thermometer *113*
Plants 192, *192*, 195
Plasma 22
Plastics 130, 134, 138, 219
Platinum *168*, 170, 172, *173*
Plutonium 14, 18, *18*, 20
Polarized light 214
Polar winds 114
Political science 228
Pollution 182
Polonium 150
Polycarbonate 141
Polyesters 141
Polyethylene 134, *140*, 141
Polymer 214
Polymer chemistry 136
Polymerization 134
Polymers *see* Plastics
Polypropylene 141
Polystyrene 141
Polytetrafluorethylene 141
Polythene *see* Polyethylene

Polyurethane 141
Pond skater *217*
Potassium 150, 178, 179
Potential difference 69, *76*, 214
Potential energy 26
Poulsen, Valdemar 108
Power 28, 70, 214
Power station 17, *82*
Power stroke *125*
Praseodymium 150
Precession *38*, 39
Precious metals 169–173, *168–173*
Precipitate 214
Pressure 214
Primary cells 67
Primary colours *54*, 56, 62, *63*
Printing, colour 57
Prism 43, *215*
Processing, chemical 132
Processor 98
Promethium 150
Protactinium 150
Proteins *192*
Proton 11 *12*, 13, 214
Psychology 228
PTFE *see* Polytetrafluorethylene
PVC 141
Pryometer *113*
Pyromethium 150

Q

Qualitative analysis 136
Quanta 45
Quantitative analysis 136
Quantum theory 214

R

Radar *44*
Radiant energy 28, *29*
Radiant heat *see* Infra-red radiation
Radiation 12, 18, 21, 43, 49, 184, *185*
Radiation, heat 118, 214
Radiator 117
Radio 84–89, *84–89*
Radioactivity 12, 25, 214
Radio Direction and Ranging

see Radar
Radio-isotopes 15
Radiotelephony 88
Radio waves 44, *44*, 56, 57–63, *57*, *58*, *63*, 84, *84*, *86*, 87, *183*, 185
Radium *13*, 150, 180
Radon 150, 190
Rain *123*, 194, *194*
Rainbow 54
Ramjet *126*, 127
Rare earths *199*
Rare gases 190, *190*, *191*, *198*
Rayon *139*
Rays 43
Reaction 214
Recording *see* Sound recording
Red light 55
Reduction 134, 159, 214
Reflection 45, *46*, *47*, 55
Refraction 45, 46, *47*, 215
Refractory metals 176, *177*
Refrigerator 128
Relativity 17, 40–41, *40*
Residual magnetism 82
Resistance 69, *69*, *76*, *88*, 215
Resistor 70
Resonance (acoustic) 109
Resonant frequency 88
Respiration 186
Retardation 34
Rhenium 150
Rhodium 150, 170, 172
Rocket *26*
Rocket engine *23*, 127
Rock-salt 179
Rolling mill *163*
Röntgen 49
Rotation 37
Rotor *see* Armature
Rubidium 150, 178
Ruby laser *50*
Rust 188
Ruthenium 150, 172
Rutherford, Ernest 11, 14, 16
Rutherfordium 150

S

Salt *15*, 179, 215
Samarium 151
Satellite 37, *59*
Saturated solution 215
Savery, Thomas 122
Scandium 151
Scanning 60, 62, 90, 91, 93, *94*, *95*
Science 228
Sea level 182
Seawater 180
Selenium 151
Selenium cell *70*
Self-excitation effect 82
Semi-conductor 66, 215
Series motor 76
Shunt motor 76
Siderite 157
Silicon 151
Silicon chips *101*
Silicones 141
Silver 151, 169, *169*, 170
Silver plating 166
Si units 229
Slip ring 79, *80*
Snow *123*, *194*
Soap 127
Social sciences 228
Sociology 228
Sodium *15*, 55, 152, *153*, 178, 179, *179*
Sodium hydroxide *see* Caustic soda
Solders 175, 216
Solenoid 72, *78*, 216
Solid *11*, 216
Solid-fuel rocket 127, *127*
Solvent 214
Sonic boom *105*
Sound 60, 102–109, *102–109*
Sound recording 106–109, *106–109*
Sound wave *58*, *61*, *72*, *73*, 102, *104*
Spacecraft *33*
Space engine *23*
Space rocket *26*
Specific gravity 223
Specific heat capacity 114, 216
Spectroscope 216
Spectrum: of colour 54, 216; electromagnetic 43–47, *42*, *44–47*
Speech 104
Spalerite *see* Zinc blende
Spinning body 37
Split ring *81*
Spring tide *25*
Stable atom 14, *14*
Stainless steel *161*, 167
Stars 190
Static electricity 64, 216
Stator 76, *77*
Steam 120
Steam engine *27*, *116*
Steam locomotive *27*
Steam turbine 124
Steel 188
Stereo recording 106, *107*
Sterling silver *172*
Strain 216
Stratosphere *183*, 184
Street lighting 55, 178
Stress 216
Strong interaction 25
Strontium *151*, 152, 180, 181
Structural chemistry 136
Stylus 106, *107*
Submarine 18
Sudbury ore 164, 166, 168
Sulphur 152
Sulphur burner *132*
Sulphur dioxide 134, *134*, 182
Sulphuric acid *132*, *134*, 136, *161*, 195
Sun 112
Sunlight *see* White light
Superconductivity 129, 216
Supersonic speed *105*
Surface tension 216
Suspension 216
Sync pulse *61*, 62, 93
Synthetic fibres 141

T

Talc 180
Tantalum 152
Tape recording 106, 108, *109*
Technerium 152
Telephone 72
Television 56, *56*, 87, 90–95, *90–95*; broadcasting 60–63, *60–63*
Tellurium 152
Temperature 110–122, 128, 184, 216
Tension *37*
Terbium 152
Terylene 141
Thallium 152

Thermal energy 28
Thermocouple 216
Thermodynamics 128
Thermometer 98, 112, *112, 113*
Thermometer 216
Thermonuclear power 21, 23, 52
Thermoplastics 138, 140, 141
Thermosetting plastic 138, 140
Thermostat *116*, 216
Thomson, Joseph John 16
Thorium 152
Thrust *126*, 127
Thulium 152
Tides *25*
Time 40
Tin 152, 174
Tinstone see Cassiterite
Titanium 152, 177, *177*
Torque 76
Trade winds 114
Transformer *82, 83, 84*, 216
Transistor *89*
Transmitter *57*, 60, *63*, 88
Transmutation 14
Transparency 43
Transverse wave 102
Trifid nebula *196*
Triode *84, 86*, 87, 90, 225
Tritium *20*, 21
Tropopause 184
Troposphere 184
Tuned circuit 86, 88, *88*, 89, *89*
Tungsten 152, 176
Tuning *109*
Tuning-fork *102*, 104
Turbine 123, *126*
Turbofan engine *126, 180*
Turbojet engine *126*
Turboprop engine *126*, 127

Two-stroke cycle 124

U

UHF (ultra high frequency) 61
UHF picture 62
Ultrasonics 225
Ultraviolet radiation 43, 184, 216
Universal force 24
Uranium 11, 12, *13*, 14, 16, 18, *18*, 20, *20*, 21, 22, 152

V

Vacuum 122, 123, *177*, 218
Vacuum flask *207*
Valency *14*, 218
Valve see Diode; Triode
Vanadium 153
Van der Graaf generator *219*
Vapour 218
Variable capacitor *89*
Vector diagram *35*
Velocity 34, *35, 40*
Vibration 102, *102, 104*, 106, *106, 109*
Viscose rayon *139*
Vitamins 136
Volta, Alessandro *66*
Voltage 66, *88*, 90
Voltaic cell *66*
Voltmeter *76*
Volumetric analysis 136

W

Water *14*, 113, 118, *118, 123*, 182, 194, *194*

Water wave 102
Watt 28, 218
Watt, James 28, 123
Wavelength 44, *44*, 51, 218
Waves 87, 102
Waxed cylinders 108
Weak interaction 25
Weather picture of Europe *184*
Weaving *173*
Weight 32, 218
Weightlessness 32
Welding *187*, 197, 201
Westerlies 114
Wheel movement 33
White light *42*, 55, *55*
Whittle, Frank 125
Wind *114, 123*, 184
Wire *165*
Wolfram see Tungsten
Work 28
Wrought iron 160

X

Xenon 153, 190
X-rays 43, *48*, 49, *181*, 218

Y

Yellow *55*, 57
Yttrium 153

Z

Zero 112, 113
Zinc 153, 174
Zinc blende *174*, 175
Zirconium 153
Zoology 228
Zworykin, Vladimir 91, *95*